CALIFORNIA:

THE GREAT EXCEPTION

by Carey McWilliams

Foreword by Lewis H. Lapham

UNIVERSITY OF CALIFORNIA PRESS

Berkeley · Los Angeles · London

University of California Press
Berkeley and Los Angeles, California

University of California Press, Ltd.
London, England

First California Paperback Printing 1999

Library of Congress Cataloging-in-Publication Data

McWilliams, Carey, 1905–1980
    California, the great exception / by Carey McWilliams.
      p.   cm.
    Originally published: New York : Current Books, 1949.
    Includes index.
    ISBN 0-520-21893-0 (alk. paper)
     1. California—Miscellanea.   I. Title.
F861.M25   1999
979.4—dc21   98-42930
                                                        CIP

Printed in the United States of America
1 2 3 4 5 6 7 8 9

The paper used in this publication is both acid-free and totally
chlorine-free (TCF). It meets the minimum requirements of
American Standard for Information Sciences—Permanence of
Paper for Printed Library Materials, ANSI Z39.48-1984. ∞

For a fine fellow—

JERRY ROSS McWILLIAMS

TABLE OF CONTENTS

Contents

CONTENTS

ix

Contents

FOREWORD

CAREY McWILLIAMS published *California: The Great Exception* in 1949, and the California that he describes is the one that I remember as a boy growing up in San Francisco during the excitements of the Second World War. He writes about a port that was then the busiest on the American Pacific coast, and I can still see the crowd of ships, aircraft carriers as well as cargo vessels, riding at anchor in the bay; he mentions the names of once prominent citizens, and I can see the people in the streets dressed as characters from a script by Raymond Chandler. The men wore hats and double-breasted suits; the women wore fur and high-heeled shoes. Together they danced to the music of Cole Porter and the Andrew Sisters. Veronica Lake was in love with Alan Ladd, a computer was a giant robot confined to the realm of science fiction, San Jose was somewhere vaguely south on El Camino Real, a dusty farm town where Mexicans wrapped in blankets dozed in the shade of the eucalyptus trees, and if Herb Caen had been asked to guess what was meant by the word silicon, he most likely would have said something about an insect repellent or a Chinese tailor who had figured a new way to make silk shirts.

The changes brought about by the passage of the last fifty years have been many and various, but none of them seem to me more remarkable than McWilliams' understanding of what hasn't changed. Although most of his statistics have faded and nearly all of his projections have proved too modest, his book remains current because it proceeds from his appreciation of California as a temperament, a metaphor, a turn of mind. Like Minerva springing full-blown from the head of Zeus, California emerged full-blown from the myth of Golconda, its origin coincident in 1848 with the discovery of gold at Sutter's Mill and with what McWilliams calls "the magic equation" found in the American river along with the fortune-bearing gravel and the miraculous sand. The gold deposits ranged across an escarpment roughly 300 miles long and 50 miles wide, present at depths varying from a few inches to a few hundred feet, and for twenty years they offered the chance of a nabob's riches to anybody who cared to come and dig. Never before or since in the annals of the American dream did its promise of equal opportunity prove so unmistakably true—room enough and gold enough for everybody on the sunset horizon of the bountiful American frontier.

The gold rush attracted expectant capitalists from everywhere in the world—not only other Americans from the eastern states but also Frenchmen, Chinese, Mexicans, Italians, Irishmen, Germans, Dutchmen, Swedes—all of them optimists, most of them young and male, few of them burdened with the luggage of civilization and its discontents. Because they arrived at more or less the same moment (San Francisco enlarging its population from 800 in 1848 to 30,000 in 1851) they got off to a more or less even start in the new country, a country without an established social order, without government, without law, tradition, system, prior claimants. The volatility of the abruptly formed mass produced McWilliams' equation, "gold equals energy," which in turn prompted the all but instantaneous creation of something new under the sun. Citing the testimony of Bayard Taylor, an early traveler to San Francisco who likened the city to "the magic seed of the Indian juggler, which grew, bloomed, blossomed and bore fruit before the eyes of his spectator," McWilliams observes that "In California the lights went on all at once, in a blaze, and they never have dimmed."

The sentence provides the theme to which he adds a set of well-informed variations, deriving from the state's exceptional geography its equally exceptional history, agricultural enterprise and political practice. Evolving outside the continuum of a gradually extended frontier, California was admitted to the union in 1850, and because it was excused from a term of apprenticeship as a territory, the Californians acquired the habit of making up their own rules. By 1864 they had taken $100 million in gold from the public domain without paying a dollar in taxes, and with their newfound wealth they made a commercial empire—iron foundries, shipping companies, eventually banks and railroads—that owed nothing of its existence to old ideas, settled monopolies, eastern money. As fond of luxury as they were of gambling and dancing, the inhabitants consumed "seven bottles of champagne to every one consumed by Bostonians," and the motley character of their society—plural, cosmopolitan, tolerant and unstable—guaranteed a freedom of movement and encouraged, or at least didn't frown upon, a freedom of thought.

It is no accident that California over the last 150 years has provided the country with so many of its new directions, most obviously in the computer and entertainment industries but also by way of its enthusiasm for Reaganism, environmental ballot initiatives, smudge pots, sexual experiment, aircraft design, hybrid fruits and vegetables. The belief that wealth follows from a run of luck fosters among the Californians (now as in 1848), a willingness to deal the cards, take the chance, entertain the proposition from the gentleman wearing the mismatched boots or the lady with the parrots. Who knows? Maybe one of them will bring rain.

Nor is it surprising that the state continues to attract adventurous spirits from all points of both the moral and geopolitical compass. McWilliams thought it amazing that between 1940 and 1948 no fewer than 3 million new people had come into the state, and when he looked at the estimates that posted California's population in the year 2,000 at a number as high as twenty million, he was reluctant to credit the projection. By 1995 the population had reached thirty-two million, and the annual income from its agriculture, which in 1949 McWilliams thought extravagant at $2.3 million, had swelled to what he undoubtedly would have regarded as the inconceivable sum of $2.6 billion.

But although he might have been surprised by the prodigious fact—in the same way that he was surprised by the size and weight of a California squash or a California artichoke—he would not have been surprised by the even more prodigious dynamic. If he understood

California as a magic equation, he also knew it as a plural narrative, not one story but many stories, none of them simple and all of them about the search for promised fortune—for gold and land and water, for a new identity, an old hunting ground, a government contract or a string of horses, for something seen in the play of sunlight on a canyon wall or in a drift of rain through tall trees. What remained constant was the dreaming energy of the California mind, its delight in metaphor and its wish to believe in what isn't there, about the future as a work of the imagination and a past that came and went as abruptly as last year's movie set or yesterday's snow.

Lewis H. Lapham
November 12, 1998
New York

California:

THE GREAT EXCEPTION

ON UNDERSTANDING CALIFORNIA

◇◇

> I MET *a Californian who would*
> *Talk California a state so blessed,*
> *He said, in climate none had ever died there*
> *A natural death.*
>
> —ROBERT FROST

WITH CALIFORNIA noisily celebrating three centennials—
the discovery of gold (1848); the adoption of the first state constitution (1849); and admission to the Union (1850)—a question
first raised a hundred years ago and never really answered has
acquired a new urgency: Is there really a state called California
or is all this boastful talk?—Is this centennial only ballyhoo,—
a hoax, a fraud, a preposterous imposition? The question has
bobbed up again because people have always been dubious about
a state whose name is something of a hoax. No one knows, of
course, the origin of the word "California" or whence it came or
what it means. It first appears as "Califerne" in the *Song of
Roland* and was probably borrowed from the Persian, *Kari-i-farn*,
"the mountain of paradise." Deeply encrusted with myth and
legend, the name is historically associated with a hoax, Marco
Polo's mention of a fabulous isle "near the coast of Asia" which
apparently no one ever saw or mapped or set foot upon. Although
its derivation is unknown, California has a meaning which is as
clear today as when the word stood for a place not yet discovered.
It is the symbol of the mountain of paradise; the fabulous isle;
the dream garden of beautiful black Amazons off the Asia coast;
"the good country"—the Zion—of which man has ever dreamed.
Naturally, people have always been wary of this great golden
dream, this highly improbable state; this symbol of a cruel illusion.

CALIFORNIA: *The Great Exception*

Like all exceptional realities, the image of California has been distorted in the mirror of the commonplace. It is hard to believe in this fair young land, whose knees the wild oats wrap in gold, whose tawny hills bleed their purple wine—because there has always been something about it that has incited hyperbole, that has made for exaggeration. The stories that have come out of California in the last hundred years are almost as improbable and preposterous as the tale from which the state gets its name. Although the exceptional always incites disbelief, it comes to be accepted as perfectly normal by the initiated; and thus a problem in communication arises as different standards of credence emerge. Like Alice to whom so many out-of-the-way things had happened that she had begun to think that very few things indeed were impossible, the Californians have acquired a manner of speaking that arouses ridicule. The failure of understanding that has resulted is based on the difficulty of avoiding the hyperbolic in describing a reality that at first seems weirdly out of scale, off balance, and full of fanciful distortion. For there is a golden haze over the land—the dust of gold is in the air—and the atmosphere is magical and mirrors many tricks, deceptions, and wondrous visions.

Not recognizing this danger, those who have written about California fall into two general categories: the skeptics who, in retrospect, have been made to look ludicrously gullible; and the liars and boasters who have been confounded by the fulfillment of their dizziest predictions. Hinton Helper, one of the first of the skeptics, bitterly denounced the California of the gold rush as "an ugly cheat," vastly overrated and greatly overdone; a state where "nothing is as it should be" and every event seems "as momentous and unaccountable as the wonderful exploits of Aladdin's genii." The fact that it didn't rain between April and November struck Helper as being symbolic of the deceitfulness and perversity of a state whose every outward form was somehow a snare and delusion. But, amusingly enough, this "ugly cheat," this most improbable state, has always made the skeptics look silly and, as a symbol, has lost none of its potency. That 3,000,000 people have trooped into California in the last eight years shows, in the most alarming manner, that the golden legends still flourish. But, since nothing is yet quite what it should be in California, a section of American opinion still refuses to take seriously a land which seems to distort fact but in which the real distortion is in nature. The implacable Helper complained that California was "already a pandemonium" in 1848 and pandemonium it remains. Any doubts on this score were removed by the amusing antics that took place when, in an effort to ape a solemnity they did not feel, the citizens of California undertook the first ceremonial observances of the state's triple centennial.

On the morning of January 24, 1948, thousands of automobiles began to converge on the sleepy little town of Coloma (population, 300), on the south fork of the American River, to celebrate the centennial of James Marshall's discovery of gold in California. Two narrow, winding foothill roads are the only means of reaching Coloma. The sun was hardly up before both roads were jam-packed with cars, bumper-to-bumper, with traffic paralyzed for fifteen miles. As the cars inched their way toward Coloma, the people laughed and shouted, and ran up and down the line of march exchanging drinks and greetings. By noon 75,000 people were surging through the streets of tiny Coloma. Long before noon, however, the improvised booths were emptied of souvenirs, the food supply was exhausted, and the hotel was a shambles. With no place to stay overnight, visitors began to push their way out of town toward nightfall, although many, in despair of the traffic, curled up in their cars and went to sleep.

While the celebration lasted, Coloma was in the grip of a second gold rush which brought ten times the number of people who had assembled there, a hundred years ago, when the place was a gold camp of 10,000 population. For two days, Coloma was again a fabulous boom town: prices zoomed; stores and shops were stripped of merchandise; and the competition for parking and standing room was phenomenal. By the morning of January 26th, the crowds disappeared as suddenly as they had arrived, and the dazed residents of Coloma, still groggy, began to sweep up the littered streets and remove the tattered bunting from the store-fronts. Coloma's second gold rush had come and gone and now

Coloma, and California, had crossed the threshold into the second century of the state's meteoric rise to fame and power.

The confusion, incongruity, and disorder of the Coloma celebration, inaugurating the centennial of the discovery of gold, are symbolic of the still on-rushing, swiftly-paced tempo of events in California. At the famous Philadelphia centennial of 1876, the visitors could at least pause and reflect upon the course of events and the significance of the occasion. The crowds were large but they moved slowly, swinging canes and parasols, taking their time, enjoying a new sense of maturity. But the crowds that descended on Coloma were in a hurry, pushing their way into Coloma from Red Dog and Gouge Eye, from Hangtown and Lotus, Slug Gulch and Poker Flat, and speeding back along the winding roads once the celebration was over. They had not come to pause and reflect but to have a drink and be on their way. There was no pause in California's observance of its centennial, for, if anything, the tempo of events had been stepped-up with the passage of time. This was not just a centennial, but a lark, an outing, a split second's interruption in the busy, heedless lives of the Californians.

In fact it was quite apparent in Coloma, on January 24th, that California was not prepared to celebrate its centennial. A hundred years had passed, to be sure, but the Californians had to work awfully hard to bring off the illusion of lapsed time. The local male residents of Coloma donned flannel shirts and sported whiskers, and the ladies of the Mother Lode appeared in the bonnets and calico gowns of yesteryear. But no one was fooled by this innocent deception; everyone knew that the celebration was a hoax. Although California has more than its share of poets, no one was asked to write a Centennial Ode, for how could any poet invest this jamboree, this awkward traffic snarl, this rip-roaring clambake, with overtones of solemnity and high purpose? The calendars said that a hundred years had passed but, in terms of symbolic truth, the celebration was premature.

Just as California cannot properly celebrate its centennial, so the time has not yet arrived for a real summing-up; one cannot, as yet, properly place California in the American scheme of things. The gold rush is still on, and everything remains topsy-turvy. The analyst of California is like a navigator who is trying to chart a course in a storm: the instruments will not work; the landmarks are lost; and the maps make little sense. The last eight years have been, in fact, the most dynamic years in the history of this most dynamic state. No, the time has not come to strike a balance for the California enterprise. There is still too much commotion—too much noise and movement and turmoil.

What I have attempted, therefore, is in the nature of an essay in understanding—a guide to an understanding of California. The following chapters might be described as the notes, the working papers, of a California journalist; the summation, not of California, but of my effort to understand California. There is, however, a theme which runs through the following pages—that California is "the great exception" among the American states. There is also a purpose, namely, to isolate the peculiar dynamics underlying California's remarkable expansion.

CALIFORNIA—THERE SHE GOES!

Du RING THE war Californians were aware vaguely of a phenomenal increase in population. From time to time officials made speeches heralding the dawn of a *New West* and occasional headlines hinted at a great post-war expansion. But every one was too preoccupied with the war itself to give much thought to what was happening in the state. In fact the full "shock of recognition" did not come until August, 1949, when the Bureau of the Census released a report on population shifts for the period from April 1, 1940, to July 1, 1947. If the nation was amazed to learn that, in this period, California had gained 3,000,000 new residents, it would be fair to characterize the reaction in California as a curiously ambivalent mixture of pride, consternation, and dismay.

No other state in the American union has ever shown a volume of increase through migration even remotely approaching the gain that California has registered in the last eight years. In fact the gain is so large as to represent a substantial redistribution of the population of the United States. In this period, more people moved to California than were living in Los Angeles County before the war. Historically we have learned to think of the westward movement of population; but what we do not realize is that for the last forty years the westward movement of population has been primarily a movement to California. Nor has this movement stopped. The experts now forecast that California will show an additional gain of 2,650,000 in the 1950's; that it has not yet reached the mid-point in its growth. It is expected that 20,000,000 people will eventually reside within its boundaries. Although it is now generally agreed that the West has somehow "come of age" with the centennial of the discovery of gold in California, the nation has still failed to comprehend the meaning of the continued mass migration to California. In a later chapter I will deal with the phenomenon of migration to the west coast—a subject in itself; here I merely want to describe what has happened in California in the last eight years and to point out some of the consequences and implications.

TIPPING THE SCALES

In the last eight years, the three west coast states led the nation in population growth. Their combined population increased by 3,981,000 or 40.9 per cent, and now stands at 13,714,000. During this period, California passed Illinois and Ohio in population and edged closer to Pennsylvania, the second largest state in the Union. According to later unofficial estimates, California's gain was 3,123,-613, Oregon's 536,316, and Washington's 751,809. Percentage-wise, California's increase was 45.2 per cent, Oregon's 49.2 per cent, and Washington's 43.3 per cent. As many people migrated to Oregon in the last eight years as in the entire first century following the arrival of Lewis and Clark at the mouth of the Columbia River. During the same period, as many people settled in California as were living in the state at the end of the first World War.

It is extremely difficult to assimilate the significance of a population shift of this magnitude. The very magnitude of the increase has obscured the point that there is essentially nothing surprising about the facts. California has not "boomed" in the last eight years; it has continued to grow at a more or less *normal* rate. In the last hundred years, the population of California has registered an increase of about 44.6 per cent per decade or approximately 3.8 per cent per year. It should be noted also that, in accordance with the "law of growth," California is still a very young state whose area is virtually limitless in comparison with its present population. California's present population density, per square mile of arable land (not counting mountains, desert, and forest) is only one-eighth that of Massachusetts, the first state to be settled. If California continues to follow what population experts call "the law of growth," it will expand at an almost constant, but gradually declining, rate for the next two or three decades. It may be inferred, therefore, as Dr. William A. Spurr has pointed out, that California's population will ultimately exceed twice its present level, or 20 millions. This is really "news," and it is something for the nation to ponder.

What the nation does not realize is that population shifts have a dual significance: one region's gain must necessarily represent another region's loss. Over a period of time, therefore, a shift in population can bring about significant changes in interregional relationships which in turn can have far-reaching social, economic, and political implications. In the last eight years, nine states actually lost population: Arkansas, Kentucky, Mississippi, Oklahoma, West Virginia, Nebraska, North Dakota, South Dakota, and Montana. It is also important to contrast *rates* of increase for the same period: 6.1 per cent for Pennsylvania, 5 per cent for New York, 6.3 per cent for Illinois, 45.2 per cent for California. Texas, which had approximately the same population as California in 1940, has failed to keep pace with its western rival (it showed a 10.7 per cent increase for the same period). Seven states in the west north Central Division—Minnesota, Iowa, Missouri, North Dakota, South Dakota, Nebraska, and Kansas,—failed to keep pace with the average rate of growth for the nation and the southern states also fell below the national average. In point of fact, therefore, only the west coast states were well above the national average rate of growth.

The significance of these figures is obscured by the phrase "westward shift in population." The West's increase in population is highly concentrated; not all states in the region have shown substantial increases. Colorado showed a slight increase, but Idaho lost 40,000 in population, and Montana 65,000. Thus the postwar problems of these states are almost diametrically unlike those of the west coast states. If Arizona, Nevada, and Utah are excepted, then it can be said that the Inter-Mountain states either held their own or showed losses in population. With an overall average civilian gain for the nation of 10.6 per cent, only the three west coast states showed gains substantially above this figure. Michigan, Indiana, Ohio, Delaware, New Jersey, and Connecticut scored gains above the national average but these gains only ranged from 11 to 17 per cent. Finally it should be noted that the spectacular increase for the three west coast states represented an increase in urban population. New industries were, for the first time, the magnet that drew 4,000,000 people to these states in an eight-year period.

What does this shift of population mean in political terms? Since the number of representatives in Congress is fixed by law at 435, it is quite apparent that in 1950 some states will have to forfeit representation in order to accommodate the three west coast states. This in turn will change the regional balance of power within the nation. California will probably be given a minimum of six additional seats in Congress and two seats will have to be allotted to Oregon and Washington; a total of eight. These eight seats will have to be deducted from the representation of other states. Even states which are increasing in population but which have failed to keep pace with the leading states, Missouri is an example, will be affected by this redistribution. Thus New York, with 45 seats in Congress, will probably have to yield three seats to the west coast states.

Eight seats in Congress represent a substantial political increment; but, remember, in regional terms, this means that an equal number of seats must be deducted from other states. Since urban areas of the west coast have been receiving the bulk of the new migration, this shift also means that urban industrial areas in the three west coast states will gain politically at the expense of rural areas. It should also be kept in mind that California's influence in the electoral college and in the national political conventions of the two major parties will be substantially increased. There can be little doubt, therefore, that, as the *New York Times* recently observed, "California can no longer be thought of merely as the Land of Sunshine. Politically and economically, she tips the national balance westward."

CHANGE IS CUMULATIVE

Although taken aback by the Census Report of August, 1949, national opinion, as reflected in editorial comment, tended to rationalize the shift in terms that made for a feeling of complacency and self-satisfaction on the part of the older states. "After all," these editorial writers seemed to say, "we still hold the power; the nation's population will soon become stabilized; and, in the last analysis, natural limitations will place a brake on California's expansion." In general the attitude was that of a friendly uncle taking a certain measure of satisfaction in the achievements of a west coast nephew; but, in this case, "uncle" has failed to grasp the true significance of his nephew's phenomenal growth. Since the rapid prior growth of the west coast failed to work any profound shift in social, economic, or political power, "uncle" has been prone to see little basic significance in the developments of the last eight years. But there is this all-important difference: it is now the growth of new industries that is attracting population to the west coast, and the growth of industrial power, particularly the upsurge of California, has profound national significance. What "uncle" has failed to note is the appearance of a new set of population dynamics on the west coast.

As the always observant Richard L. Neuberger has pointed out, "the most hopeful factor in Oregon's economic situation is now the immense new population in the State of California." It is this aspect of the matter that "uncle" has largely overlooked. In the past, almost every article produced in the Northwest had to be shipped eastward across the continent to the major national markets at freight rates which were always discriminatory and often prohibitive. Now the Northwest has discovered that it has a promising and ever-expanding new market at its doorstep. California, with 10,000,000 people, represents quite a market. In fact more people now live in California than in all of New England. "Whenever we examine in detail the shipment of Oregon products," writes Bernard Goldhammer, economist for the Bonneville Power Administration, "we inevitably discover that a preponderance goes to California. This applies to agricultural commodities, to lumber, to aluminum, to cheese, to nearly any item one can enumerate." Rubber and textile factories in California purchase the product of a new rayon plant in Eugene, Oregon, and the remarkable expansion of the furniture industry in Los Angeles nowadays provides an excellent market for Oregon timber. As automobile manufacturers establish assembly plants in California, and experiment with plants to manufacture parts, the aluminum industry of the Northwest suddenly assumes a new significance. According to Van Beuren Stanberry, a special economist for the Department of Commerce, "California has now become a major market for the products of the Northwest. Oregon once had to ship lumber and cheese 2,000 miles to find a market of 10,000,000 consumers. Now such a market lies at the end of the 700-mile Shasta route of the Southern Pacific out of Portland."

It is not by chance, therefore, that the volume of north-south train, bus, and airline passenger traffic on the west coast has begun to exceed in importance the volume of the east-west traffic. Nor is it surprising that the consumer goods and service industries along the coast have outstripped new manufacturing. Skeptics will point to the fact that the residents of the west coast are deluding themselves; that they are only active because they are "taking in each other's washing." But this can hardly obscure the fact that an entirely new dynamic has come into being with the phenomenal population increase in California—an increase which is almost certain to continue, at present rates, for the next two decades. "Uncle" is still a rich man, full of wisdom and cunning, heavy of girth, with a fat bank account; but he had better take a second close look at his ambitious nephew.

WESTCHESTER, THE WAR BABY

Statistics on population growth, although they may impress the expert, fail to convey the reality of what has happened in California in the last eight years. Just what does it mean to dump 3,000,000 people into a state, even a state as large as California, in the brief period of eight years? Although the absorptive capacity of the state is still very great, the latest rush of people to California has produced an impact not unlike that of the gold rush a hundred years ago. Actually thirty times as many people have come to California, in the last eight years, as came during the gold rush decade. The impact of this latest migration has been all the greater by reason of the fact that the war migrants surged into already crowded cities and into a limited number of these cities, mainly San Francisco, Oakland, San Diego, and Los Angeles. Since most of the migrants came to Los Angeles, it is to this city that one must turn for illustrations of the new type of community that has come into being; the community which represents the modern day equivalent of the "gold camp" of 1848. Westchester, "the fastest growing community in the United States," is perhaps the most interesting.

In 1940 Westchester was merely a name on the maps for a large, vacant area near the Los Angeles Municipal Airport, green in the rainy season, brown in the summer, of gently rolling slopes and level plains planted to lima beans. In 1941 there were only 17 widely scattered homes in the entire area; today 30,000 people live in Westchester. Everything about Westchester is new and shiny: its streets, its homes, its growing shopping center, its schools. Only within the last year has it begun to emerge from its camp-like, squatter phase. In 1948 precisely 5,492 homes, most of which sold for about $7,000, were built in Westchester and 8,000 additional homes are planned or under construction at the present time. Here, on the plains, a good-sized city has come into being. Although its development was almost wholly unplanned, by some miracle Westchester has the appearance of a fairly well-planned community. It is trim and neat and painfully, incredibly new. As cities go, it is about the newest thing in California. It is as though some one had waved a magic wand and a city had suddenly appeared. As might be imagined, the city that is there today where once were fields of lima beans and wild mustard, has about it the air of unreality that one associates with movie sets and other miracles of improvisation; but Westchester is quite real; it is not an illusion.

The settlers of Westchester, the pioneers of '48, were war workers who wanted homes near the aircraft factories. Since homes had to be built somewhere, for war workers, this seemed to be an ideal place. At the outset, no one thought of Westchester as a community, much less as a city; it was just a wartime improvisation, a "camp." Many of the settlers were much too busy to think of planning a community and, besides, they were not sure that they intended to stay in California. But it was not long before people began to say that they "lived in Westchester." At some point, it began to occur to an ever-increasing number of people that a new community had been born. This consciousness of community identity is, indeed, a strange thing. Six homes, a dozen homes, two dozen, do not make a community; even a hundred homes will not always make a community. Community consciousness is not necessarily a function of size; it is more closely related, perhaps, to such factors as time and place. In the case of Westchester, everyone arrived about the same time, under approximately the same circumstances, and built or bought much the same kind of homes. The area was just sufficiently removed from other community-centered areas to set it apart, to give it an impetus toward self-recognition and a sense of identity. Whatever the cause, this collection of homes, bungalows, and cottages began to emerge as a community within a year after the first war migrants moved in.

The population of Westchester is as "young" as the community is "new." The adult population, for example, is highly concentrated in the 30-to-34 age bracket. About 75 per cent of the men are veterans of World War II. Not only is the bulk of the population "young" but there are practically no "old people" in Westchester; and this is not quite the same thing. Most of the residents are in the middle of the middle class with extremes of both wealth and poverty being largely absent. For the most part, the men work in the skilled trades, the professions, civil service, and in manufacturing plants; few of the women work outside the home. Practically everyone in Westchester (90 per cent of the residents) own or are purchasing their homes. The school population, of course, is as "young" as the adult population: only 49 per cent of the children have yet reached the age of school enrollment, a circumstance which has created a great interest in kindergartens and nursery schools. Unlike a "mining camp" of 1848, Westchester is a remarkably homogeneous community, a factor which probably accounts for the rapid growth of community consciousness. Here is a community made up of people remarkably similar in age, background, income, and interest; a community with an unusual interest in schools, playgrounds, and recreational centers because of the "abnormal" number of teen-age children. "Our children," as one Westchester housewife has said, "have not yet reached the age of delinquency and we do not intend to have any delinquency in Westchester."

This statement throws a clear light on at least one aspect of the widespread, post-war social ferment in California. The amount of lethargy in community attitudes probably increases in direct ratio to the age of the community. To change a pattern, to change anything in fact, seems to be more difficult than to establish a new pattern, and particularly with Americans, a notoriously impatient and restive people. Thus, by a paradox, the lack of planning created in Westchester the challenge to plan; the newness of the community, the "youth" of its population, and its homogeneity, provided the dynamics which made planning possible. It has been said that newcomers in California are reluctant to develop an interest in community affairs; but in Westchester the interest in general civic affairs is unusually great. Approximately 57 per cent of the adults are registered voters, a somewhat higher percentage than for Los Angeles as a whole.

This "ferment of newness" is shown in other matters. Not enough churches have yet been built to take care of the religious needs of the community. By necessity, therefore, the existing churches have had to share their facilities; the Jewish congregation uses the Baptist Church and most of the churches exchange pastors. Inter-faith activities of all kinds have been stimulated, and the existing churches have come to occupy a new relationship to the community. In the absence of other facilities, churches have become the equivalent of a town hall or city council. No one factor, of course, explains the absence of a warring sectarianism in Westchester; it has come about as a result of a peculiar combination of social circumstances.

Here, then, is an eight-year-old city of 30,000 inhabitants with no local fire or police stations, and without emergency hospital facilities. Although an integral part of Los Angeles, there is no direct telephone line to the area so that the residents must pay a toll charge on all calls. The "city library" is a building about the size of a box car, and the elementary schools are a collection of hastily thrown together bungalows. Never formally planned, the streets of Westchester are a jumble of unrelated numbering, criss-crossing, and sharp turns; only the oldest inhabitants can find their way about with ease. Although a shopping center is developing, this city of 30,000 inhabitants is, at the last report, without a barber shop. Yet, despite these omissions, inconveniences, and limitations, Westchester is going ahead, raising money to build a town hall, seeking, by a variety of devices, to improve community services.

CALIFORNIA'S GROWING PAINS

California, the giant adolescent, has been outgrowing its governmental clothes, now, for a hundred years. The first state constitution was itself an improvisation; and, from that time to the present, governmental services have lagged far behind population growth. Other states have gone through this phase too, but California has never emerged from it. It is this fact which underlies the notorious lack of social and political equilibrium in California. The state is always off balance, stretching itself precariously, improvising, seeking to run the rapids of periodic tidal waves of migration. Right now it is trying to negotiate the latest and the most dangerous of these recurrent "rapids." The tensions created by the constant lag between government services and population growth can best be appraised in light of the fact that, since 1940, California has added to its population the equivalent of the entire population of the state of Virginia. During the last seven years, enough people have come into California every month to make up a city of 40,000 population. Just what this means in terms of a constant lag between services and needs can be shown by a quick survey in a few key areas of government.

School enrollment in California was 14 per cent greater in 1947 than in 1945; in fact the kindergarten enrollment was up by 28 per cent. In 1948, California faced the task of providing school facilities for 100,000 more children than in 1947. Currently Los Angeles, with 260 average births per day, needs 30 new schools and will have to add about 30 additional schools in the next five years. In 1948, 27,000 children were forced into part-time attendance in the Los Angeles schools because of the shortage of facilities; in one year the enrollment shot up by 19,800. Since many of the wartime migrants were young people, birth rates have been rising rapidly and the state faces a real school crisis between 1955 and 1960. Needless to say, this situation has created a shortage of everything related to the schools, including teachers. If present trends continue, Los Angeles alone will be short 8,000 teachers in the elementary schools by 1955. If every man and woman graduating from every school of education in the state between now and 1955 were to get a job in the Los Angeles school system, there would still be a shortage of teachers. The pressure is greatest, of course, in the elementary schools; but it will soon be felt all along the line.

The same "crisis" appears in other fields. Today, Los Angeles is the third largest metropolitan area in the nation, second only to New York and Chicago. Its population has jumped 35 per cent— more than 1,000,000 people—since 1940. With new residents coming in at the phenomenal rate of 16,000 a month, it goes without saying that housing and hospital facilities will be greatly overtaxed. Third in size, Los Angeles ranks 18th in the number of hospital beds per person. It must build 52 new hospitals in the next 20 years. The burden on correctional institutions and institutions for the mentally ill has been proportionately great. With the largest veteran population of any city in the nation—some 715,000 veterans reside in Los Angeles—the local Veterans Administration has been fighting desperately to keep abreast of the avalanche of new claims and new cases. Over 407,000 new telephone installations were made in Los Angeles in a three-year period: more than the company had made in the eight busiest pre-war years. Library facilities have lagged far behind population growth. Traffic plans have become obsolete before they have emerged from blueprints. Community chest drives have fallen far short of their stated goals. Sewer facilities in one community after the other have been overtaxed to the point of creating grave public health hazards.

Planning, in such a state as California, has suddenly taken on an entirely new dimension. For the plain fact is that no calculus exists by which needs can be fully anticipated in California. Other communities can project a population curve and, with fair accuracy, anticipate needs twenty and thirty years in advance; but it would be a brave man, indeed, who would undertake to chart California's growth for the next decade. There are too many unpredictable factors; too many variable elements.

Aside from the inherent difficulties of planning in California, the nature of the state's population growth creates special resistances to large-scale planning. Even at the limping pace at which facilities have been expanded, and they have never kept abreast of current needs, governmental costs have skyrocketed. For the average Californian, the expenses of state, city, county, and district government have increased four times since 1910. Pointing to the upward curve of governmental expenses, and comparing this rate with other states, reactionary interests consistently confuse the voters and minimize the need for a rapid expansion of government facilities. The fact is, of course, that comparisons with other states are wholly misleading.

The vehicle in which California is attempting to run the current rapids is laughably ancient and obsolete. The state constitution is a monstrous patchwork of 340 pages, the second longest state constitution. Most of its provisions are utterly outdated, and have been for many years. "Born in a boom," the first state constitution was amended and re-amended between 1849 and 1879 as the population increased more than seventeen times. In 1879, when the second constitution was adopted, the population was 864,694; it is today 10,031,000. At last count, the 1879 constitution had been amended 235 times and every year from 50 to 70 constitutional amendments are proposed. Since 1879, of course, all the powerful organizations have gotten their particular pet schemes, their "sacred cows," written into the state constitution; so that the adoption of a new state constitution presents a well-nigh insoluble political problem. With the consequence that, for nearly a century, California has been dragging along in a one-horse shay.

Adopted only 20 years ago, the present charter of the City of Los Angeles is today almost as obsolete as the state constitution. It, too, is a ponderous document of 295 pages, containing 513 sections; to know it well is a life's work. Since its adoption, the charter has been amended 260 times. The county of Los Angeles, of course, is a governmental monstrosity. Within the county are 45 independent municipal governments, varying in size from 1,000 population to 2,000,000 population; from a few square miles in area to 470 square miles (Los Angeles proper). Within the county 500,000 people live in unincorporated areas and there are small "pockets" of county territory juxtaposed with incorporated areas. Within the City of Los Angeles are dozens of "conscious provinces," such as Hollywood and Eagle Rock, which continue to think of themselves as separate municipalities.

Californians, of course, are fascinated by facts and figures showing the state's phenomenal growth and yet, on another side of their minds, they are disturbed and even repelled by these same figures. They want the state to grow, and yet they don't want it to grow. They like the idea of growth and expansion, but withdraw from the practical implications. This ambivalence is so acute that it often results in paralysis, a suspension of the thinking faculty, a form of civic hypnosis. Each wave of migration is regarded with fear and trembling, and the wave next before the last invariably comes up with the idea that the latest arrivals are "inferior" to those who came at an earlier date.[1] Without exception, these rationalizations are always based on editorial fancy rather than fact.

With the lifting of gas rationing in August, 1945, the press of Southern California carried stories with such headlines as "Migrant Workers Flock Homeward,"[2] and "Exodus East Continues."[3] One could detect in these stories a note of quiet jubilation as the older residents demonstrated a familiar willingness to speed the parting guest. One day after Japan surrendered, 417 cars loaded with furniture, bedsprings, mattresses, baggage, children, dogs, and goats passed through the Arizona border station on the backward swing to Oklahoma and Arkansas. For weeks the exodus continued as the newspapers carried joyous stories that "The 'Grapes of Wrath'

[1] *See:* editorial, Los Angeles *Daily News*, May 27, 1948.
[2] Los Angeles *Times*, August 16, 1945.
[3] *Ibid.*, August 26, 1945.

Folks have reversed their field with the sudden advent of peace and there is now an ever-growing exodus from Southern California." What a relief! One could almost hear the official sigh of pleasure as the migrants turned eastward.

But one year later almost to the day the border patrol reported that 130,000 people had entered California from Arizona in a single month, their noses and radiators pointed toward the promised land. Consternation immediately spread through California's officialdom. By September 1946, the Mayor of Los Angeles was urging that "steps" be taken to slow up, preferably to reverse, the influx of migrants into Los Angeles. As a matter of fact, the westward movement had started within three months after the exodus began—the Okies and Arkies had merely gone "back home" for a vacation. By December 1945, the by-now-familiar returning movement was well under way and the headlines read: "State Lures Record Influx of Visitors"; [4] and "Swelling Migrant Tide Poses Perplexing Issues." [5] By mid-1946, to judge from the howls, wails, and shrieks of protest that came from California officials and the state's short-memoried press, one would have thought that California was being inundated with a swarm of locusts, not people.

This astonishing ambivalence, so amusing to watch, consistently undercuts any attempt to plan for the well-being of Californians, present and future. The unconscious rejection of the migrants paralyzes the need to plan for their assimilation and adjustment. The Californians never quite believe in their good fortune; it appears to be real enough but then, again, it could be an illusion. Formerly Californians believed in attracting migrants; but the initiative has long since passed to the migrants. It is the migrants who are planning to come to California; not California that is planning to receive them.

PLANNING BY INDIRECTION

With all these inhibitions of the planning function, how then does it happen that the influx of 3,000,000 did not produce a state

[4] *Ibid.,* December 18, 1945.
[5] *Ibid.,* December 20, 1945.

of chaos? There are many answers to this question. For one thing, California has space to burn. The City of Los Angeles has the largest land area of any city in America: 44 miles by 25 miles; enough land to support a population of between eight and ten million people. The county of Los Angeles, with 4,038 square miles, is about the size of the state of Connecticut; New York is only one-tenth as large. In terms of space, Los Angeles has been able to absorb an enormous increase in population with the minimum inconvenience. People simply fill up the vacant spaces.

The spread-out character of Los Angeles, plus the volume and velocity of migration, has resulted in a natural and, from many points of view, a highly desirable dispersion of population. Industries are widely scattered in Los Angeles. For the most part, the war-time growth of Los Angeles has taken place on the periphery of the community, rather than at the center. In some respects, if this development had been planned, it could not have been more desirable. By an accident, therefore, Los Angeles has become the first modern widely decentralized industrial city in America. For, with the growth taking place in the peripheral areas, the city has found it more convenient to decentralize services and facilities than to attempt a new integration from the center. As fast as new areas have developed, the chain stores, the department stores, and the drive-in markets have chased after the people, setting up new shopping districts and establishing new neighborhood centers. With more automobiles per capita than any city in America, and with the worst rapid transit system of any city, Los Angeles was almost ideally prepared for a decentralization which it did not plan but from which it will profit in the future. The demonstrated unresponsiveness of these peripheral areas to directives issuing from the center, and their exaggerated sense of self-importance, have also been factors in the pull of services from the center to the margins.

One of the great problems in Los Angeles is that many of the city's institutions have not adjusted to the decentralized pattern of the city. The metropolitan daily newspapers have simply resigned from the task of multiple community reporting and have fallen back increasingly on county-wide, national, and international news.

On the other hand, some 250 separate newspapers have sprung up all over Greater Los Angeles, to reflect the interest and news of particular neighborhoods and communities. The new Los Angeles *Independent*, formed by merging a number of neighborhood shopping papers, is now attempting to get out 12 separate editions, each of which will carry the news of a particular locality as well as city-wide and county-wide news and events of national and international interest. The *newness* of sections of Los Angeles has created opportunities for which planners have dreamed for many years. San Fernando Valley, not so many years ago a "rural" section of Los Angeles County, today has a population of 350,000 and, by the end of the century, may well have 1,000,000 residents. In other areas, planners have only begun to plan for the "satellite" city, the decentralized community of from 35,000 to 50,000, with its own services, residences, and industries. But Los Angeles is already made up of a series of "satellite" cities, all unplanned, but for which some planner will doubtless claim credit in the future.

Another clue to the success with which California has assimilated 3,000,000 new residents in eight years, is to be found in the character of the migrants. It is not an easy task to absorb, in less than a decade, a population substantially equivalent to the entire state of Iowa. California's migrants, however, represent a selection rather than a cross-section of the American population. Many of them are veterans; perhaps 250,000 veterans have settled in Los Angeles alone since the end of the war. They are young people, active, in their best working years; 45 per cent, for example, are between 15 and 34 years of age. On the whole, they are much younger than the resident population, the median age of which, in 1940, was four years older than the average for the nation. Often referred to by the California press as "undesirable," the war migrants show a higher proportion of college graduates than is to be found in California. The number of high school graduates, among the migrants, is considerably higher than the average in the states from which the migrants have come. Three-fourths of them come from points west of the Mississippi River. Although many of them are "unskilled," there is a high percentage of skilled workers included in the total.

For the most part, then, the migration of the last eight years has been made up of people who have quickly and easily adjusted to the conditions of their new life in California. The same characteristics of the total migrant group can be found, for example, in the large wartime influx of Negroes to California. Today Los Angeles County has the third largest concentration of Negroes outside the southern states, with perhaps 320,000 Negroes now residing in the county. There are, however, certain sections of the migrant population that present a special problem, particularly the "senior citizens." In 1940 there were 10,000,000 people in the United States over 65 years of age, of whom 750,000 resided in California and, of this group, 325,000 resided in Los Angeles County. In the same year, 6.8 per cent of the nation's population was over 65; but the percentage in Los Angeles was 8.5 and may now be close to 10 per cent. One-fourth of Los Angeles County's "senior citizens," those over 65, are receiving some form of public assistance.

This, then, is California in 1948, a century after the gold rush: still growing rapidly, still the pace-setter, falling all over itself, stumbling pell-mell to greatness without knowing the way, bursting at its every seam. Today it has 10,000,000 residents; tomorrow it may have 20,000,000. California is not another American state: it is a revolution within the states. It is tipping the scales of the nation's interest and wealth and population to the West, toward the Pacific. The nation needs to understand this tawny tiger by the western sea, and to understand this tiger all the rules must be laid to one side. All the copybook maxims must be forgotten. California is no ordinary state; it is an anomaly, a freak, the great exception among the American states.

[3]

THE MAGIC EQUATION

◇◇◇

IF ASKED to name the most important respect in which California differs from the other forty-seven states, I would say that the difference consists in the fact that California has not grown or evolved so much as it has been hurtled forward, rocket-fashion, by a series of chain-reaction explosions. The rhythm of the state's development is unlike that of the other states, and the basic explanation is to be found in a set of peculiar and highly exceptional dynamics. The existence of these underlying dynamics accounts for the tempo of social change, the foreshortening of economic processes, the speed of development. Europeans have long marveled at the driving force, the "restless energy," of America; but it is only in California that this energy is coeval with statehood. Elsewhere the tempo of development was slow at first, and gradually accelerated as energy accumulated. But in California the lights went on all at once, in a blaze, and they have never been dimmed. It was, of course, the discovery of gold that got California off to a flying start, and set in motion its chain-reaction, explosive, self-generating pattern of development. Not gold alone, but the magic equation "gold-equals-energy," is the key to the California puzzle.

The discovery of gold in California, which *Harper's Weekly* characterized in 1859 as "perhaps the most significant, if not the most important event of the present century connected with America," was providentially timed. Nine days before the Treaty of Guadalupe Hidalgo was signed, gold was discovered in California. None of the negotiators of the treaty, of course, were aware of the discovery when the treaty was signed. The discovery came, therefore, just in time to direct and accelerate the swelling tide of American emigration toward the Far West. It populated California overnight, and accomplished in a few months what Imperial Spain had been unable to accomplish in three centuries. It made San Francisco not only a major port but the capital, for fifty years, of the western empire. It drew the first transcontinental railroad directly to San Francisco. And, above all, it got California off to a flying start, decades ahead of the other western states. To appreciate the kinetic effects of the discovery of gold in California, one must recognize that the California gold rush is unique in the annals of gold discoveries, and that it brought into being an utterly unique mining frontier. "California appears to be the only place," writes Dr. John W. Caughey, "where a rush for gold was made to serve as the base for an ever-widening superstructure of attainment." In just what particulars, therefore, was this gold rush unique?

POOR MAN'S GOLD RUSH

The California gold rush was unique, first of all, in that the discovery of gold in California coincided with a revolution in the means of transportation and communication which made possible a mass migration from all points on the compass. The number of people participating in the prior gold rushes of the eighteenth century was not large. News of the California discovery was diffused, as W. P. Murrell has pointed out, "as widely as the 19th-century newspapers could spread it." Furthermore, all the resources of modern ocean transport were available for all who cared to go to California. The California gold rush was, therefore, the first gold rush to set in motion a world-wide mass migration.

Furthermore, the California gold rush was the first, and to date the last, poor man's gold rush in history. The gold-fields were located in California on the public domain. Every miner in California was a trespasser on the public domain and nearly every ounce of gold produced in the state belonged to the federal government. But, in the confusion of the period, the American military commander "prudently decided that he would permit all to work freely" in the diggings. In sixteen years of "free mining" in California, over $100,000,000 was taken from the public domain without a dollar's revenue passing to the federal treasury. There were no squatters, no prior claimants to the gold lands in California; and, since there were no regulations, it was quite impossible for anyone to acquire title to a mining claim other than by holding it and working it. This made for an extraordinarily *rapid* development, and a truly amazing democracy in production. Elsewhere, in New South Wales, in Africa, in Siberia, gold discoveries were quickly monopolized, either by prior claimants, the government, or by the circumstance that the deposits were concentrated in particular areas. Hence none of the other gold rushes had anything like the stimulating effect that the discovery of gold produced in California.

Gold deposits were found in California over an area nearly 300 miles in length, from 40 to 100 miles in width, and at depths varying from a thin veneer to 300 feet. "The extent of workable deposits was so great," writes Dr. Ralph H. Brown, "that conflicts over claims were extremely rare." Not only was there ground enough for all to work, but the widely diffused nature of the deposits made it impossible for any group to secure a monopoly. There is a wealth of evidence to show that, during the first decade at least, "there was room and gold for all." Here was an equality of opportunity almost unmatched in history for, since most of the miners faced a situation "that was new to all alike," none of them had a monopoly of knowledge on methods or locations. Few could conquer with Pizarro or sail with Drake, but the California gold rush was the great adventure for the common man. During the term of the gold rush, say from 1848 to 1873, the equality of opportunity typical of the frontier was, as Murrell has said, "never better exemplified" than in California.

Even more than in the case of an agricultural frontier, this exceptional mining frontier made for a real equality of fortune. Not only was there ground enough and to spare for all to work, but the average yield was high: perhaps an ounce of gold per day for every miner. Since the price of gold was fixed by world competition, and not by local production, increased gold output in California did not lower wages and the California miners universally regarded the day's "wash" as the equivalent of a wage. Not only were wages high, but a vast number of miners made individual fortunes (and, of course, promptly lost them). Four hundred men, working on the American River in 1849, produced an average daily yield of from $30,000 to $50,000 in gold. Governor Mason reported that he knew of two men who had produced $17,000 in gold in seven days and of a woman who had "washed" $2,125 in 46 days. Within a few years, as Dr. Caughey has pointed out, the "Californians came to have more money per capita in hand and in circulation than any other people anywhere."

Since there was no "law of mines" in 1848, the California miners adopted their own rules and regulations in which they were careful to safeguard the equality of opportunity which had prevailed at the outset. California was preeminently the home of what has been called "the small mines claim" system. The rules adopted in the California camps carefully emphasized the policy of "one miner, one claim"; barred slavery from the mines; and based rights, not on ownership, which could not be established, but on prior discovery and use. These same rules also narrowly limited the size of mining claims. Later, the California miners successfully resisted, for some years, a series of measures by which the federal government sought to convey fee titles to mining claims. These measures, the miners contended, would make for monopoly. Hence in California, unlike the other western mining states, the free miner remained, at least until 1873 or later, the foundation of the whole system.

The first mining in California, of course, was of the placer variety; and throughout the gold rush placer mining was of paramount importance. As late as 1870, ninety per cent of the state's gold was derived from placers and this percentage was placed as high as 70 per cent in 1880. By its very nature, placer mining makes for democracy-in-production. In proportion to total output, the number of producing units is always greatest during the placer mining stage, and this was notably true in California where the placer deposits were widely distributed. Placer mining, in other words, bears about the same relation to mining generally that homesteading does to farming. One man's chance of making a strike is about as good as the next man's. Furthermore, the individual miner, with his pan or rocker, is the most economical unit of production, since little advantage flows from combination of claims or from larger units of production. In the California field it was almost impossible to employ wage-labor, as the temptation to prospect always offset the advantages of regular employment. Since wages were fixed by the average daily wash, and not by competition, there was little advantage to be gained by employing labor.

One might think that the abundance of money in circulation in California would have forced interest rates down; but exactly the contrary was the case. During the first decade of the gold rush, interest rates stood at 3 to 5 per cent *per month* and were often higher. One reason, of course, for these rates was that there were no banks in California; in fact, incorporated commercial banks did not come into existence until the middle 1860's. Then, again, with the uncertainty of titles, no one cared to loan money on mining ventures. The twin factors of high wages and high interest rates account for the fact that during the term of the gold rush itself, that is, for a quarter century, most mining operations in California were conducted by individual miners, mining partnerships, and "small parties" of miners. The mining partnership was so important, in fact, that Murrell states that in a sense it "replaced the family as the basic social unit." Until about 1860 and perhaps even later, the greater part of the immense amount of money invested in water companies, mines, and mining claims in California belonged to "parties in the mines." Mining was conducted, in other words, upon the basis of direct personal ownership and responsibility. In the other western mining areas, however, the incorporated company, almost from the outset, became the standard and well-nigh universal unit of production. But from 1848 to the present time there has been very little "eastern money" invested in California mining. Not until 1876 was any California mining stock quoted in the daily San Francisco market reports, and as late as 1874 a mining journal reported that "it is a significant fact that the majority of good quartz mines of the State are in private hands and pay well enough in themselves, without the necessity of the owners having recourse to stock-jobbing operations."

Metals are always exploited, of course, in a descending order of price: first gold, then silver, then copper, and, finally, lead, zinc, and iron. From California, where the American mining frontier began, the mining wave rolled eastward to Nevada and from there spread throughout the inter-mountain West. In these areas, however, silver and copper were of greater importance than gold and, as a consequence, placer mining quickly gave way to quartz mining. Comstock Lode, the second great strike in the West, was quite unlike the California placer camps. "In place of the independent miner of the California placers," writes Murrell, "its typical figures were the skilled wage-earning miners." By the time the silver boom was on in the West, Congress had begun to assert federal control of mineral resources on the public domain. And by then, too, mining law, as it had developed after 1848, greatly favored large claims and capitalist methods of exploitation. In fact, the nature of quartz mining, which required heavy capital investment, dictated this change in the method of exploitation. It is important to note, therefore, that the Coeur d'Alene, Cripple Creek, Butte, and Bisbee "strikes" all came *after* the mining industry had become highly capitalistic and *subsequent* to the adoption of a federal law of mines. In 1866 Congress enacted legislation which permitted titles to be acquired to mining claims; and later enactments, in 1870 and 1872, enlarged the size of claims which could be acquired, permitted the use of proxy claimants, and otherwise modified the California "small claims" system.

This change in the character of the mining frontier, as it rolled eastward, had enormous social consequences. The transition reflected, of course, the difference between placer and quartz mining. Little capital was required in the former; large sums in the latter. The placer deposits were widely distributed in California; the quartz deposits were concentrated in a few major districts in the inter-mountain West. Placer mining was based on the small claim and the rule of one miner, one claim; quartz mining gave rise to the large claim, multiple filings, and the growth of monopoly. "As mining developed," writes Murrell, "it became more and more dependent, first through ditches (sluice and hydraulic mining), then through increasingly elaborate machinery, on capital, and more and more demanded exploitation in large units. . . . In the new mining industry to which he had pointed the way the miner, once the freest of Americans, was relegated to a subsidiary place."

Unlike the discovery of gold in California, the staggering copper-and-silver production of the inter-mountain West acted as a drain on the wealth and resources of the areas in which the mines were located. In California gold had created a reservoir of local wealth which was used in agriculture, trade, commerce, banking, shipping, and industry, and, incidentally, to develop western mining; but the great mineral wealth of the inter-mountain states was largely syphoned off to non-resident owners and stockholders. Miners lived in camps in California; in the West they lived in company towns. From 1848 to 1948, California produced about two billion dollars in gold; but Montana, Idaho, Utah, and Arizona have produced five times this sum in copper alone, without this production having anything like the stimulating effect that the discovery of gold produced in California.

As a consequence of this difference between two types of mining frontiers, the inter-mountain West, America's last frontier, was, paradoxically, the *first* region to feel the full impact of the new social integration brought about by large capital combinations, monopolization of resources, and the use of large-scale units of machinery. The inter-mountain West, in other words, stepped directly and immediately from its frontier phase into large-scale industrial production; from the creekbed claims to the tunnels of Butte. A similar change eventually took place in California, but the transition in the inter-mountain West was much more sudden and violent. The western miner, once the freest of Americans, was not easily subdued and the record of his resistance is to be read in the bloody chronicles of the Coeur d'Alene, Cripple Creek, the Ludlow Massacre, and the Bisbee deportations. The great strike which developed in the silver mines of Leadville in 1878 has been characterized by Dr. Frederic L. Paxson as "one of the great forerunners of economic clash" in America; one of the first, if not the first, "modern" industrial strike in American history. To a very large degree, California escaped the debilitating, socially disastrous, and economically ruinous effects which the emergence of this new type of mining produced in the inter-mountain West.

CALIFORNIA: THE MINING ENTREPRENEUR

It was from California that men, money, and machinery poured into the western mining areas. Wherever he went, the California miner carried his newly acquired lore of mining, his rapidly developing technology, and his mining rules and regulations which became the cornerstone of the American law of mines. The discovery of gold not only brought a new state into being in California, but it was this state which largely peopled and built up the intermountain mining districts. Many of these districts were California colonies. California, writes one mining historian, was "the Mother of these Pacific States and Territories. . . . What England is to the world, what the New England states have been to the East, California has been and still is to the country west of the Great Plains. Her people have swept in successive waves over every adjacent district from Durango to the Yellowstone." How was it, asks Dr. Rodman W. Paul, that a young western state, itself not two decades old, was able to play the role of entrepreneur in this vast new territory? His answer is this: "California was *unlike* any previous commonwealth that had existed west of the Alleghenies."

Its "unlikeness" consists in the richness and diversity of its resources which the discovery of gold unlocked, not slowly and tortuously, but overnight, in two decades. The discovery of gold brought about, as Dr. Caughey has said, "a stimulation of commerce that was the most intense the American frontier had seen." This stimulation took the form of a chain reaction which affected every segment of the economy. Overnight a large population had gathered in a frontier, isolated state, and the presence of this population, so plentifully supplied with purchasing power, created an enormous market for goods and services of all kinds. The factors of time and distance were such, furthermore, as to overcome the disadvantages and handicaps of industrial and agricultural production in a frontier area. Miners needed tools, supplies, food, lumber, transportation facilities, wagons, leather; in fact, there was hardly anything they didn't need and couldn't pay for.

The eastward expansion of the mining frontier began in 1859

with the discovery of the Comstock Lode. By this time California had made rapid strides in the development of its trade, commerce, agriculture, and industry. The California gold rush had reached its term by 1873; but the boom in western mining lasted for a much longer period. Thus the stimulating effect produced by the discovery of gold in California continued as the mining frontier expanded throughout the West. It was not only California machinery and supplies, however, that were used in the outlying mining district: a large part of the capital came from California. For fifty years or more, San Francisco was, in every sense, the mining capital of America. The business of supplying miners and mines has always been more profitable than the business of mining, witness the careers of Huntington, Stanford, and Crocker; and California was the supply center for western mining. The construction of railroads greatly stimulated western mining for it made possible the exploitation of low-grade ores and one of the principal rail lines in the West was controlled by California interests. Factors other than distance precluded the eastern industrial centers from enacting the role of supply centers for western mining. For example, the Civil War absorbed all the productive facilities of these centers, thereby giving California a headstart in the business. Furthermore, none of the other gold "rushes" had anything like California's high record of technical achievement. One could write a weighty treatise on the mining innovations, inventions, processes, and techniques that were first developed in California. California manufacturers, therefore, had a marked advantage in technology, in experience, in know-how.

In appraising the impact of the discovery of gold on the economy of California, one might say that gold production was the least significant aspect of the discovery. By 1860 the value of manufactures in California exceeded the value of gold production by twenty million dollars; in the same year California's wheat crop exceeded the value of the gold produced in the state. In 1866 there were 13 iron foundries and 30 machine shops in San Francisco, and 23 iron foundries in other parts of the state. The value of iron castings produced in San Francisco alone was about two million dollars. Virtually all of this production was related to mining. But, once mining demands began to taper off, and as eastern manufacturers began to invade the market, California had a small industrial plant, the only one of its kind in the West. What the gold rush had initiated in California in the way of an industrial plant was, therefore, given a further rocket-like forward propulsion by the expansion of the American mining frontier. It is a real historical anomaly to find a frontier state playing the role of financier, manufacturer, and supplier to other frontier states; but that is exactly what happened in the West. It should be noted, finally, that California was able to play this role, in the last analysis, because of the diversity of its resources. It was the only western state that could have supplied the needs of an expanding mining frontier.

THE "SOMETHING FOR NOTHING" BUSINESS

What has been said in the foregoing section should not obscure the very real importance of gold mining in California. The gold produced in California had value, and large amounts were produced. But the value of this gold consisted in more than its purchasing power. The *activity* that its production generated, the energies that it released, were the important considerations. The production of gold, as Thorstein Veblen once said, is the "something for nothing" business. Economically it is an utterly nonsensical and thoroughly wasteful business. The history of gold production the world over shows that the total cost of the supply of gold habitually exceeds the total value of the output by several hundred per cent. Industrially gold production is a waste; but the activity which it generates is a most powerful economic stimulant.

One can make a most impressive case in support of the point that gold production did more harm than good to the economy of California. The senseless explorations and wasteful methods used did irreparable damage to forests, farm lands, and river systems. Much of the labor that went into the production of gold was completely wasted. In the long run, most of the miners got a very small return for their labors. Indian villages in California were engulfed and destroyed by the spread of the mining frontier. Furthermore the gold produced was not valuable, in the sense that iron is valuable; for gold is only useful, writes Dr. Caughey, "for beauty and dentistry." But gold production is the incomparable stimulant to trade and business and industry, for it involves manifold activities. It is the very best economic pump-primer. For example, one flume and aqueduct constructed in northern California during the gold rush was 70 miles long, cost a million dollars to build, and its construction kept a large crew busy for a year. The production of gold created more problems for California than it solved; but it was nevertheless "the touchstone" that set California in motion toward greatness and power. From 1848 to 1860, eastern coal miners were lucky to receive a wage of $1 a day; but the average daily wage in the California mines was $3, and, for most of the period, $5 a day. This, again, is another measure of the value of gold as a pump-primer.

But by far the greatest value of gold to California was its value as a symbol. Overnight California became a world-famous name and, as a name, California meant gold. It was the discovery of gold that catapulted California into the national limelight; that increased its population 2,500 per cent in four years; that gave it statehood within two years after the discovery. A state that gets off to this sort of flying start possesses advantages that do not disappear with time and changed conditions. The tide of migration which the discovery of gold set in motion is still running strong. The world-wide publicity which the discovery gave the state is still a potent factor in its development. The plain fact is that it is quite impossible to appraise the importance of the discovery of gold in California, for the ramifications are endless. Examine any phase of California life—agriculture, labor, government, industry, social organization—and the examination inevitably involves some consideration of the importance of the discovery of gold. Nothing is more exceptional about this exceptional state than the unique combination of factors and conditions produced by the discovery of gold. Nothing quite like it has ever occurred, or is ever likely to occur again, in world history.

There is, however, a most peculiar relation between the discovery of gold and the nature of California's resources. The chain of events which the discovery set in motion was precisely of the character necessary to unlock the resources of the state. Gold was about the only resource which California possessed that was to be had, so to speak, for the asking; that one could simply take and possess. Most of California's resources are of a character which have required a high level of technology to unlock. But, the rapidity with which population increased in California due to the discovery of gold, and the prices which prevailed, brought about a rapid, large-scale development. Had it not been for gold, one can assume that these resources would have been developed step-by-step, item-by-item, instead of on the grand scale and more or less simultaneously. It takes great wealth to produce great wealth in California. If the great riches of the state had not been unlocked, so to speak, with one motion, they might have languished for decades. In other states, forced growth is used to supplement organic or natural growth; but in California forced growth is the rule, almost, one might say, a necessity of production.

Historians have a fine time in California speculating about the course that events might have taken had gold not been discovered. But one can better measure the importance of the discovery by posing another hypothetical question, namely, what would the discovery of gold have meant had the discovery been made, not in California, but in any of the other western states? It would have meant, of course, a boom followed by a bust. But California was charged with all sorts of dynamite in the form of latent or potential resources. Gold was the fuse and the spark that touched off these explosions. Hence the chain reaction effect which continues to the present time. Resources have not been developed in California on a piecemeal basis but in "wholes," as entities, and this is precisely the manner in which exceptional resources of this character should be developed. This peculiar relation between the kind of resources which existed in California and the energies which the discovery of gold released has produced the exceptional dynamics which, for a hundred years, have been propelling California forward, not by steps, but by strides; not by inches but by miles. It is not by chance, therefore, that California is the one locality in the world where a "rush for gold was made to serve as the base for an ever-widening superstructure of attainment."

The Magic Equation

THE LIGHTNING'S BOLT

That fateful bolt of lightning that Marshall had released touched off first one, then another, cache of dynamite hidden in California; and each of these explosions touched off still others. The chain reaction started in 1848 and, if you listen, you can hear, from time to time, the explosions which are still propelling California forward. This forward movement has not been steady, straight, and consistent; it has gone forward by swift turns, by self-generating spurts. California is like the mechanism of the ratchet wheel and pawl. A burst of energy, another explosion, sends the wheel spinning forward and then it locks until the next burst of energy sends it spinning again. When Californians speak of oil as "black gold" and of lettuce as "green gold," the reference to gold is more than a figure of speech. For the effect of these developments, in the explosive environment of the state, has been quite similar to the discovery of gold. Each step forward, each advance, has set in motion still another chain reaction. All the gold ever mined would amount to only a small fraction of the total oil production.

While one can marvel at the ratchet-like forward movement of California, it goes without saying that the process is searing and disruptive, making for disequilibrium and a noticeable lack of stability, producing many strange institutional maladjustments and social derangements. In California you learn to wait for the next explosion and, when it comes, you run as far and as fast as you can and then dig in until the next explosion splits the air. The process, also, imposes a strong strain on one's sense of reality, as it makes for skepticism on the part of those who have not experienced these furious forward movements. "Of all the marvelous phases of the Present," wrote Bayard Taylor, "San Francisco will most tax the belief of the Future. Its parallel was never known, and shall never be beheld again. I speak only of what I saw with my own eyes. Like the magic seed of the Indian juggler, which grew, blossomed and bore fruit before the eyes of his spectators, San Francisco seemed to have accomplished in a day the growth of half a century."

Taylor, it will be noted, was not quite sure that he believed what he had seen, any more than J. S. Hittell could suppress some skepticism in reporting the existence in Santa Barbara of a grape plant with a trunk 15 inches in diameter, its branches supported by an arbor 114 feet long and 78 feet wide, with an annual yield of three or four tons of grapes; of a nugget, discovered in November 1854, that weighed 145 pounds; of squashes weighing 210 pounds; of a beet that weighed 118 pounds; of a turnip that tipped the scales at 26 pounds; or of pear trees that grew 10 feet in one year and plum trees that shot up 12 feet in as many months. All he could say, when cross-examined about these prodigies, and it is all I can say, is that California, "her plants, her quadrupeds, her birds and her fishes, are different from those of other countries." For plants and trees, vegetables and fruits, grow in California like cities grow; and the same peculiar dynamics seem to be universally at work. California *is* really different.

[4]

THE MUSTANG COLT

◇◇

> I'LL *bet my money on the mustang colt,*
> *Will anybody bet on the grey?*
> —CAMPTOWN RACES

THE WESTWARD movement of population in America began rather slowly; the penetration of the Allegheny ridges was, at the time, a formidable undertaking. After this first barrier had been crossed, the wave of settlement rushed westward to the edge of the Great Plains, hesitated for a moment, and then, with the discovery of gold, broke westward for California. Up to this point, certain familiar phases had characterized the expansion of the frontier: exploration, conquest of the Indians, settlement, the birth of institutions, the marking-off of boundaries, territorial government, and, finally, statehood. But this pattern was broken once the wave of settlement reached California. Just as the velocity of the westward movement increased as it neared the Pacific, so the process of frontier settlement seems to have been speeded-up. California did not go through the various phases of this process, serial-fashion, but stepped immediately from frontier to statehood. The deviation from the norm, in this case, consisted in more than a change of pace. It was, in every respect, quite unique. The fore-shortening of the process of frontier settlement in California, moreover, has had important latter-day consequences.

CALIFORNIA: THE OUTSIDER

The prime factor which accelerated the westward movement of population, once the edge of the Great Plains had been reached, was the discovery of gold in California. The population movement then "leap-frogged" over the inter-mountain west to the coast. This meant, of course, that the supply bases, the starting points, of the westward movement receded in space and time. The magnet of gold pulled people *through* the inter-mountain region and thereby delayed its settlement. Thus for the first time a break, a gap, occurred in the familiar process of frontier expansion. This break left a great void between the settlements in California and the nearest eastern outposts of settlement. California was settled, therefore, out of sequence, and, as a result, was isolated.

Frontiers always develop in isolation but the isolation of California was a special case. In 1848 California was 2,000 miles removed from the western edge of the frontier. The two highest mountain ranges in the United States and hundreds of miles of desert and rocky wasteland lay between California and the nearest eastern settlements. The shortest routes across this vast expanse of territory were the Gila Trail, leading to San Diego, and the Spanish Trail, terminating in Los Angeles. But the discovery of gold in northern California shifted most of the traffic to the Truckee Route, which was even more forbidding and difficult than the southern trails. The sea approaches to California, around the Horn and across the Isthmus, were equally difficult and even more time-consuming.

The early isolation of California under American rule merely repeated its isolation under Spanish and Mexican rule. Both by sea and land, California was about as remote from the centers of population in Mexico as it was from those in the United States. As the most remote outpost of Spain, California, as E. C. Semple pointed out, was "distinctly *a peripheral state with the usual tendencies toward defection.*" Remoteness had made it a difficult province to populate and the absence of warlike nomadic Indians had always kept the local garrisons at a minimum. Both Spain and Mexico found that California was a difficult province to administer; witness the fact that California had more revolutions than the rest of the borderland settlements combined. There was always, as Blackmar noted, a strong sentiment for independence in California.

Unlike the other frontier states, with the exception of Oregon and Washington, California was both a sea and land frontier; and, unlike these states, it had both a sea and land frontier in relation to Mexico, Central, and South America. Along the Pacific Coast, from Alaska to the tip of South America, the mountains rise directly from the sea, but, in California, east of the Coast Range, is the great Central Valley. This valley is the largest arable area *west* of the Sierra Nevadas and, in relation to the Pacific, it is the American counter-weight to the great plains of eastern China. Just as California occupies, therefore, a somewhat anomalous geographic relation to the rest of the country, so it occupies a similarly unique relation to the Pacific Basin. Measured in terms of comfort, money, and time, California was actually nearer to China and South America, prior to 1869, than it was to the Mississippi.

The isolation of California was both in quality and degree quite unlike that which prevailed elsewhere on the American frontier. Hittell stated this difference most concisely when he wrote that "Other new states were in substance merely the expansion of the outer boundaries of older states; but California was essentially a colony and developed as a distinct and for the time being a disconnected organization." Essentially California developed "outside" the framework, the continuum, of the American frontier. The difference is that between a child raised in the home of his parents, with relatives and familiar surroundings, and the child taken from his home at an early age and brought up in a remote and different environment. Under American, as under Mexican rule, California was "distinctly a peripheral state with the usual tendencies toward defection."

THE THIRTY-FIRST STAR

If ever a state was admitted to the Union under highly exceptional circumstances California was that state. The story of California's admission to the Union has been told many times, but, in this section, I want to emphasize certain aspects of the story which have had a direct bearing on latter-day trends and developments.

Had it not been for the discovery of gold, events in California might have taken a slower, a more casual and "natural" course. Not only did the discovery of gold stimulate the desire on the part of the federal government to consolidate the conquest as quickly as possible but the influx of population made consolidation imperative. Delay in extending the revenue laws to California had resulted in large losses to the federal government and these losses promised to assume ever larger proportions. The situation was aptly summarized by Judge Peter H. Burnett, who later became the first governor of California: "The discovery of the rich and exhaustless gold mines of California produced a singular state of things in this community, unparalleled, perhaps, in the annals of mankind. We have here in our midst a mixed mass of human beings from every part of the wide earth, of different habits, manners, customs, and opinions, all, however, impelled onward by the same feverish desire of fortune-making. But, perfectly anomalous as may be the state of our population, the state of our government is still more unprecedented and alarming. *We are in fact without government,*—a commercial, civilized, and wealthy people, without law, order, or system."

In 1849 the population of California was divided into three major groupings: about 10,000 "native Californians" of Spanish-Mexican-Indian descent, concentrated in the southern counties; several hundred "old residents" who had arrived in California prior to the discovery of gold; and about 100,000 who had flocked to the state to mine for gold. Two out of every three of these newcomers were foreigners; they were mostly young men with a thirst for adventure and a taste for lawlessness; and, from every point of view, they were a most heterogeneous lot. The newcomers, for the most part, were concentrated in the northern and central portions of the state. Under these circumstances the early imposition of some system of government was imperative, if only for the reason that the Mexican-American War was still being fought in California between the newcomers and the conquered *Hispanos* who were concentrated in the southern part of the state.

The logical thing to have done would have been to create a territorial government but three considerations precluded this solution. For one thing, "a large and harassing political question" was then being debated in the United States—the question of slavery—and the parties to this controversy could not agree upon the form of territorial government for California. The balance between these opposing forces was very close in 1849; the Senate was equally divided and there was no slave state to pair with California should it be admitted as a "free" state. Moreover, the geography of California posed a peculiar problem for the state extended from the Mexican border to parallel 42 north; hence a projection of the Mason-Dixon line westward would have forced a division of the state. Division was unthinkable because the southern part of the state was inhabited by the "disaffected" Mexican element. To have cut off this portion of the state might well have stimulated an irredentist movement and, with the shadow of civil war lengthening over the land, neither side wanted to take this risk.

In the second place, the isolation of California raised very difficult administrative problems. A territorial government would have faced the same difficulties that the military commanders faced. Events were moving too swiftly in California to be dealt with by remote control. The third, and decisive factor, had to do with the attitude of the Californians themselves. Both new and old residents alike were quick to realize that the discovery of gold had made California "an apple of contention" between the "free" states and the "slave" states and, at the same time, had given California an extraordinary bargaining power. They were also quick to realize that, given the national deadlock on the slave question, the debate on territorial status might drag on indefinitely.

These considerations gave rise to an almost universal concurrence in the sentiment that "we have a question to settle for ourselves; and the sooner we do it, the better." The Californians were anxious to skip the territorial phase altogether, if this could be done, since, like all frontier people, they regarded government by remote control as an unmitigated nuisance. More than anything else, perhaps, it was the peripheral or "outside" relation of California to the rest of the country that quickened the sentiment for immediate statehood. "The people," wrote Joseph Ellison, "had an exalted conception of the importance of their state, whose gold, they claimed, had saved the impoverished east from bankruptcy." Thus it was that the Californians broke the deadlock by adopting a state constitution and applying for admission to the Union. In doing so, they were forced to act, as President Zachary Taylor complained, in a most "unprecedented" and highly "irregular" manner; and they have, by and large, been prone to act in this manner ever since.

Without any real authorization, the American military governor proceeded to issue a proclamation in which he asked the Californians to elect delegates to a constitutional convention. The purpose of this convention, as someone aptly described it, was "to make something out of nothing," that is, to improvise a government. Although elected in a most irregular manner, the 48 delegates who assembled in Monterey were a cross-section of the American people in 1849. Of the delegates, 36 were American citizens; 7 were "native Californians"; and 3 were foreigners. The 7 "native Californians"—read "Mexicans"—were all representatives of the *gente de razón* element; that is, they were people of quality, of substance. Only 2 of the 7 could speak English. Of the 36 American citizens, 22 were from the northern, 14 from the southern states. An Irishman, a Scot, and a Frenchman made up the "foreign" contingent. Occupationally, 14 were lawyers, 11 farmers, 8 merchants, 3 soldiers, and 2 printers, along with some minor occupations. This was probably the youngest body of men ever assembled in a state constitutional convention: 9 were under thirty, 23 under forty, and 12 between 40 and 50 years of age. Only 4 of the delegates were over fifty.

Most of the delegates were from out-of-state and had only recently arrived in California. Dr. Oliver Meredith Wozencraft, later appointed Indian Commissioner, had been in California only four months and Dr. William Gwin, who played a leading role in the convention, had hardly stepped off a boat in San Francisco before he was elected a delegate. Most of the delegates, in fact, were total strangers when they met in Monterey. All in all, one might say that this was a most typical "California" delegation. Representing, as they did, all sections of the Union, they demonstrated a remarkable ability to put sectional issues to one side. Differences of opinion seemed to cancel out by reason of the very diversity of opinion represented. Also the fact that they were strangers, and mostly newcomers, made it possible to start with a fresh slate;

there were no bothersome personal quarrels and ancient enmities. The point is not that they succeeded *in spite of* their differences in background but rather that this very diversity made early agreement possible.

THE DIGNITY OF LABOR

Given the state of national opinion at the time, one might have assumed that the issue of slavery would have deadlocked the convention. Yet, within a week from the day they assembled in Monterey, this weirdly assorted, haphazardly selected group of delegates had *unanimously* adopted an amendment affirming that "neither slavery nor involuntary servitude, unless for punishment of crimes, shall ever be tolerated in this state." Even the delegates were dumbfounded by the ease with which they had reached agreement on this crucial issue. By what miraculous alchemy had the bitter sectional prejudices of the America of 1849 been so quickly dissolved in California? That these prejudices had been "dissolved," rather than set aside, is shown by the fact that several of the delegates, but recently arrived from slave states, indicated their opposition to slavery at the convention.

This strange alchemy was to be found in two elements which were quite unique in the California situation. California was "outside" the slavery controversy, not only by reason of its geographical isolation, but because of the composition of the mass influx from the states. Prior to this time the North and the South had engaged in a bitter and active competition to colonize the western territories with elements of definite loyalty to one side or the other. The movement had been from "free" state to free territory; from "slave" state to slave territory. The nature of this pattern of settlement had merely extended the national division westward. The mass migration to California was really the first heterogeneous mass migration to the West. People came to California, not with the thought of "saving" it for the North or the South, but to make a fortune in the gold fields. Moreover the large foreign element in the population was neutral on the issue of slavery. If all the Southerners had settled in the southern part of the state, and the Northerners in the north, a different situation might have arisen; but both elements were inextricably intermingled in California. The circumstances created a strong predisposition to keep California "outside" the slavery controversy. Albert Sidney Johnston, of later Confederate fame, expressed this sentiment when he said that "Here in California there should be peace. Strife here would not be North against South, but neighbor against neighbor, and no one can imagine the horrors that would ensue."

But the decisive factor had to do with the discovery of gold, for, as Bayard Taylor had written, with capital letters, "Mining had made LABOR RESPECTABLE in California." In a sense, labor had always been "respectable" in America, that is, the labor of yeoman farmers, of artisans, of craftsmen. But labor in the modern sense was not respectable in 1849. More than one observer has pointed out that the Americans did not want "to work for" someone else; in fact, this unwillingness had been a powerful factor in the westward movement. But, in the gold fields of California, "labor" had suddenly acquired a new dimension. The bulk of the miners did not think of themselves as businessmen or industrialists. They thought of themselves as "workmen," or, as they said, "miners." The size of the earnings which could be made with a pick-and-shovel in California had something to do with this new status of labor; but there was another factor involved. It was precisely because so few of the "miners" were really "miners," because so few of them had been "workmen" before they came to California, that they wanted to emphasize the new dignity of labor. Many of these "miners" were former merchants, lawyers, storekeepers, clerks, and artisans, and, as such, they resented identification with a class which, elsewhere, had lacked status. Yet the facts compelled them to acknowledge that they were, in fact, "miners," that is, they worked in the diggings, they had calluses on their hands, and they knew what it meant to bend their backs and strain their muscles. No one who worked in the creekbeds of California could disclaim knowledge of what it meant to labor.

This new knowledge echoed in nearly every session of the strange Monterey convention like a new assertion, in American life, of the dignity of labor. "Sir," said a delegate recently arrived from Louisiana, "in the mining districts of this country we want no such competition [*that is, slave labor*]. The labor of the white man brought into competition with the labor of the Negro is always degraded. There is now a respectable and intelligent class of population in the mines; men of talent and education; men digging there in the pit with spade and pick—Do you think they would dig with the African? No, sir, they would leave this country first." "There are men of intelligence and education," said a delegate from Wisconsin, "laboring there with pick and shovel, men, who, at home, were accustomed to all the refinements of life. They are working willingly, and *they do not consider it a degradation* to engage in any department of industry. . . . But will this state of things continue, will this class of population continue to work cheerfully and willingly if you place them side by side with the Negro?"

It should be noted that miners were the dominant element, not only in the convention, but in the general population. In listing the occupations of the delegates, the historians have merely accepted the statements of the delegates as to what they did *before* they came to California. Actually the Monterey convention was a miners' convention. Nor is it possible to read the sentiments which I have quoted as chauvinistic expressions. The miners were chauvinistic in the sense that they perpetrated unnumbered humiliations and indignities upon Mexican and Chinese miners; but they were not so much "race conscious" as they were "labor conscious." As Dr. Paul S. Taylor puts it: "The debate against admission of free Negroes was not without race prejudice, but the opposition did not rest so much upon that as it was grounded upon economic fears and deep-seated philosophical objections to a rigidily stratified society." Prior to the Monterey convention, the slavery question had never been debated by delegates who so largely represented a real laboring population, and from the point of view of a society which believed that its future was to be that of a mining society. "And so it happened," writes Dr. Taylor, "that the design for California's future society was discussed not in language applied to agriculture, but rather to a developing society of gold miners." It is doubtful if any state, up to this time, had thought of its future so exclusively in terms of labor.

Both at Monterey and during the long and turgid debate in Congress over the admission of California, the point was made that California was inherently unfitted for large-scale plantation farming. Webster, in his great speech in the Senate, argued most persuasively that, since cotton could not be grown in California, and probably not in New Mexico, the question of slavery did not arise. It is, indeed, a pity that Webster is not alive today so that he might be taken on a trip through the San Joaquin Valley, where he could see some of the largest "farm factories" in America with the heaviest cotton-yield per acre in the United States. But, if his facts were wrong, his rhetoric was admirable. Although Congress hesitated and complained bitterly of California's impudence, the state was finally admitted to the Union, not on the basis of a Constitution dictated by Congress, not after a probationary period as a territory, but on its own terms, on its own initiative. The Union is an exclusive body but when a millionaire knocks on the door, you don't keep him waiting too long; you let him in.

California's gold, had it been the only factor, might not have tipped the scales in favor of admission. Ironically it was the fact that the Californians had themselves resolved the issue which Congress could not resolve that made it possible for both factions to agree on its admission to the Union. "The peculiar case of California," as its representatives argued, justified the unprecedented and irregular action which had been taken; and the fact that the Californians had adopted a constitution, uninfluenced by either faction, simplified the issue for Congress. Given the division that then existed in Congress, the logic of the *fait accompli* was unanswerable. "California," said Seward, "is *already* a State, a complete and fully appointed State. She can never be less than that."

That California skipped the "territorial phase," through which the other western states passed, has a direct bearing on the course of its development. Direct admission to the Union gave California immediate control of its own resources, unhampered by federal regulations and free of the bothersome, and time-consuming, necessity of securing congressional approval, a circumstance that long retarded the development of the western territories. It fostered a spirit of independence and a tradition of bold action. The colonial pattern, so pronounced in the other western states, is closely related to the period of territorial rule. Statehood brought California enormously important aid from the federal government in the form of land grants, gifts, and subsidies of one kind or another. But sixty-four years passed before Congress admitted Arizona and New Mexico to the Union. The contrast, here, with California is most striking and the difference in part accounts for the fact that California got off to a much more rapid start than any of the other western states.

In weighing the various factors which make for the "exceptionalism" of California, great stress must be placed on the unique circumstances surrounding its admission to the Union. For the "unprecedented" and "irregular" method by which it was admitted reflected an underlying difference in development. "In many respects," writes Joseph Ellison, "California was unique. Most new communities develop gradually; California sprang at once to full stature." Its political development, like its economic development, represents a telescoping of processes; a foreshortening of events. The Californians, in any case, have always had a lively appreciation of what it meant for the state to be directly admitted to the Union, for they have always celebrated, as a state holiday, "Admission Day," September 9th. The fact that California was the first of the eleven western states to be admitted to the Union largely accounts for the fact that it has always been a laboratory in which government has experimented with various solutions and approaches to the peculiar problems of the West. Here the government first tried out the policy of concentrating Indians on military reservations; here policy was shaped on land questions, conservation, water and power development; from California came the first mining codes, the first geologic surveys, the first irrigation districts, the first mutual water companies. The history of public policy in relation to typically western problems, in almost every instance, can be traced back to some precedent which had its origin in California.

[*49*]

THE BLANKET WITHOUT THE INDIAN

In tracing the pattern of California's exceptional status, it should be noted that "the Indian Problem" looms much larger in the other western states than in California. In Arizona, the Navajo reservation alone embraces 16,750,000 acres—an area larger than Massachusetts, Vermont, Connecticut, and Rhode Island combined; and today some 60,000 Navajos live on this reservation. In almost every western state, with the exception of California, the presence of Indian reservations has seriously complicated problems of land development, reclamation, soil conservation, and the control of watershed and forest resources. Almost any development project in these states will necessarily impinge, at some point, on the vested rights of Indians. Figures on the Indian population tell the story. In 1940 there were 19,266 Indians in Montana and Wyoming; 21,638 in Washington, Oregon, and Idaho; 20,805 in California; 11,064 in Nevada, Utah, and portions of Arizona and Idaho; 90,114 in the Southwest (New Mexico and Arizona); and 105,652 in Oklahoma. In relation to total population, such states as Nevada, Utah, Montana, Arizona, and New Mexico have a much larger Indian population than California. Or, consider the matter of Indian lands: 19,224,717 acres in Arizona; 7,153,109 acres in New Mexico; 2,739,830 acres in Washington; 1,693,160 acres in Utah; 6,454,953 acres in Montana; 1,736,794 acres in Oregon; 2,013,409 acres in Wyoming; 817,659 acres in Idaho; 666,533 acres in Colorado; and 1,127,171 acres in Nevada. Then compare these figures with 666,817 acres in California. Only Colorado has a slightly smaller Indian acreage, under the jurisdiction of the Office of Indian Affairs, than California.

To appreciate the difference, however, one must realize that in pre-Columbian times there were 130,000 Indians in California. California, it is estimated, had about 16 per cent of the aboriginal population of the United States by comparison with 5 per cent of the land area. Even at this minimum estimate, the density of Indians in California was from three to four times greater than for the nation as a whole. No one knows precisely how many Indians were living in California in 1848; but the best estimates range from 72,000 (Dr. S. F. Cook), to 100,000 (J. Ross Browne). By 1865, however, the number of Indians in California had been reduced to 23,000 and, by 1860, to 15,000. As these figures indicate, California "solved" its Indian problem by liquidating the Indians. It is true that in many other states a similar "solution" was achieved; but, as one might expect, the process of liquidation was much faster in California than elsewhere. Why was it, then, that the process was carried to such swift completion in California?

The answer is to be found, again, in the exceptional nature of the California Indian problem. "*Unlike* most frontier communities," writes Joseph Ellison, "where the advance of the white man was gradual and in a more or less straight line, in California the adventurous white settlers and miners in a short time penetrated the whole territory and partly destroyed the Indian's means of subsistence, which had never been too plentiful." The neophytes or "Mission" Indians were located along the coast, in the areas where the Spanish had settled; the "gentile" or wild Indians were concentrated in the desert, foothill, and mountainous areas to the east, directly in the line of march of the westward movement. The first Indians, therefore, that the emigrants encountered were the "wild" or nomadic Indians who, in many cases, were refugees from Spanish-Mexican rule and had excellent reasons for the hostility they exhibited. Since there were no settled Indian tribes in California, as there were in the other western states, a formal Indian frontier never existed nor could such a frontier be established. As one early day pioneer put it: "Here we have not only Indians on our frontiers, but all among us, around us, with us. There is hardly a farm house without them. And where is the line to be drawn between those who are domesticated and the frontier savages? Nowhere—it cannot be found. Our white population pervades the entire state, and Indians are with them everywhere."

Invading California from the East, the emigrants drove the Indians from their fisheries and acorn groves, destroyed the supply of fish by muddying and polluting the rivers and creeks, and, in raids on Indian villages, destroyed food supplies which the Indians had laboriously accumulated and which could not easily be replaced. The emigrants wanted, of course, the fertile valleys and rich bottom lands and from these the Indians were promptly driven. But, as the invasion spread, cattle, sheep, and hogs began to make devastating inroads on the Indians' supply of acorns, seeds, and green plants. It was only in the wastes of desert and mountain that Indians could survive, on any basis. In less than two years after statehood, California had incurred an indebtedness of over one million dollars in fighting Indians. Fighting Indians, however, was a rather profitable business in California, since the bills were always paid, sooner or later, by the federal government. In about a hundred Indian "affairs," between 1848 and 1865, over 15,000 Indians were killed. If California had been given a territorial form of government, it is quite possible that this slaughter might have been minimized. But the Californians kept badgering the federal authorities to let them handle the Indians; and, as a consequence, experienced federal officers, who knew something about Indians, were kept out of California. It was not until 1872 that Indians were permitted to file suits in the courts in California and, for many years, the testimony of an Indian was inadmissible in judicial proceedings. Under the impact of this invasion, which came overnight and with such sudden and overwhelming force, the whole fabric of Indian life was completely destroyed in California. "Never before in history," wrote Stephen Powers in a government report of 1877, "has a people been swept away with such terrible swiftness."

Since there were no settled Indian tribes, the whole question of what Indians owned and did not own in California was hopelessly confused. But the ethnologists contend that Indians had a possessory title to some 75,000,000 acres of land. In 1850 and 1851 three Indian Commissioners, appointed by the President, negotiated and executed 18 treaties with the California Indians in which the Indians relinquished all rights and claims in exchange for some 8,518,900 acres, described by metes and bounds, which the government agreed to set aside for their perpetual use and occupancy. The California legislature promptly adopted a resolution attacking the treaties and pointing out that the lands which the Indians were to receive were worth $100,000,000, containing as they did "rich veins of gold-bearing quartz." When President Fillmore submitted the treaties to the Senate on June 7, 1852, the California representatives immediately objected to their consideration and, one month later, the Senate adopted resolutions separately rejecting each of the treaties. The Indians were then forced to abandon the reservations onto which they had been moved and to seek refuge in the mountains and on the desert. After many years, the federal government finally set aside 624,000 acres to the Indians in California but these lands, as the record shows, were largely worthless. The facts, then, are these: Indians originally owned 75,000,000 acres in California; were promised 8,619,000 acres; and finally received 624,000 acres. Of the various "raw deals" which Indians received in the West, this was clearly the most flagrant.

It was because of the novelty of the Indian problem in California that the federal government first tried out, here, the policy of concentrating Indians on military reservations. According to J. Ross Browne, the reservation policy was suggested to the authorities by the fact that the Spanish had concentrated Indians in the Missions. In any case, laws were passed in 1853 setting up the first reservations in California. In selecting agents to carry out the new policy, however, the federal government seemed to show a preference, as Browne put it, for "officers experienced in the art of public speaking, and thoroughly acquainted with the prevailing system of primary elections." Cattle were purchased to the number of many thousands, ostensibly to stock the reservations, but, as the always amusing Mr. Browne reported (he was Inspector of Indian Affairs on the Pacific Coast), "the honest miners must have something to eat, and what could they have more nourishing than fat cattle?" By 1864 the reservations were practically abandoned and the Indian "problem," for all practical purposes, had been "solved."

The point to be noted about this sorry record is that California was able to extinguish Indian land titles overnight. It was never bothered, as were the other western states, with long-drawn-out negotiations with Indians, Indian agents, and congressional committees, over land titles, water rights, and similar matters. Unquestionably this factor had a great deal to do with the speed with which the economic development of California went forward by comparison with the other western states. To this day, some of these states are embroiled in the most complicated negotiations over one aspect or another of Indian rights. All one needs to do to appreciate the absence of an Indian problem in California is to glance at a map showing the location of the areas making up the 8,619,000 acres which the federal government promised the Indians. Of the 18 treaties, 15 involved large tracts of land which are today of the utmost value. Thirteen of these promised allotments were strung, like beads on a necklace, down the center of the Sacramento and San Joaquin Valleys. Had these treaties been ratified, California would have had an Indian problem the like of which cannot be imagined. But at the cost of permanent dishonor, California "solved" its Indian problem many years ago.

THE FLAG OF THE BEAR

In view of the fact that California had a population of 379,000 in 1860, the Civil War left fewer scars in this than in any other state. The Civil War was going on "back there"—"in the states"— an affair to which Californians were outsiders; interested observers, perhaps, but not participants. During the Civil War, California was rapidly accumulating, not destroying, its wealth. Unlike the eastern states, where paper money was the principal medium of exchange, specie was the only recognized currency in California. In a long controversy with the federal government, California steadily and stubbornly refused to use or to accept the new paper currency which the government issued to finance the war. By adhering to the gold standard, which was preeminently *its* standard, California was in a highly favored position. California merchants could buy in the east with depreciated legal tender and sell, in California, for gold. Trade and industry were greatly stimulated so that, at the end of the war, California was not ripe for exploitation by eastern industry and finance, but had developed its own resources to the point where it could deal with the older trade centers on a basis of equality.

The Mustang Colt

What the historians refer to as "the peculiar situation in California during the Civil War period" also had an important bearing on later developments. The peculiarity of the situation was this: from 1850 to 1861 the Democrats controlled the state government and the spokesmen for the party kept insisting that, should war ensue, California should take advantage of its remoteness and secede from the Union. It should be noted that these spokesmen were not advocating neutrality nor were they urging support for the Confederacy; what really interested them was the possibility of independence. For example, Governor Weller took the position that, in the event of war, "California should not go with the South or the North but here upon the shores of the Pacific found a mighty republic which may in the end prove the greatest of all." Other politicians echoed the same sentiment. Congressman Burch wanted California, Oregon, New Mexico, Utah, and Washington to join in forming "the youthful but vigorous Caesarian Republic of the Pacific" and proposed, as its symbol, "the flag of the bear, surrounded with the 'hydra' pointed cactus of the western wilds."

This idea, which had many adherents, was essentially feasible. California had the wealth to stand by itself; it was well protected on its eastern approaches; and it had a long seacoast with a number of excellent natural harbors. If the South should be victorious, so the argument went, it would have to respect the right of secession; if the North won, it might be too weak to undertake the reconquest of California; and, lastly, there was always the possibility of a stalemate in the war. "While politically California was a part of the union," writes Ellison, "geographically she was an isolated community separated from the central government by thousands of miles of prairie, desert, and mountains unspanned by a railroad line. This isolation naturally fostered a spirit of independence and self-reliance; a feeling that California had interests distinct from those of any other part of the Union, and a destiny of her own." It was this feeling, rather than any sentiment in favor of slavery, that gave rise to a serious secessionist movement.

Two considerations, however, doomed the enterprise. One was purely practical: the secessionists concluded that there were too few people in California to defend so long a seacoast against the possibility of attack by some foreign power which, taking advantage of the involvement of the federal government, might seek to annex California. The other saving factor had to do with the sentiment which had found expression in the Monterey convention: the sentiment against slavery. When news of the firing on Fort Sumter reached California, a majority opinion quickly rallied in support of a policy of unconditional loyalty.

This summary does not complete the story, however, for the Californians made a most skillful use of their "peculiar" situation to solve one of their most pressing problems, namely, that of transportation. Here the secessionist sentiment was used to excellent advantage. Every factor which made secession seem feasible was pointed to as a compelling reason why the government should finance a transcontinental rail line. Did the federal government want California to secede? Did it want to lose control of the gold resources of the State? Did it want to run the risk of having some foreign power, working with disaffected elements, seize this rich province?

E. C. Semple is authority for the statement that it was the federal government's fear of losing this "peripheral possession" that prompted the first survey for a rail line. "The work of construction was long postponed," he writes, "until in 1862 rumors that the people of the Pacific slope, tired of waiting for overland communications, proposed erecting an independent Republic." This rumor, coupled with the Confederate invasion of New Mexico, induced Congress to lend support to the Union Pacific project. Hence, in the midst of the war, at its darkest hour, President Lincoln, on July 1st, 1862, signed the bill which launched the first transcontinental railroad. The same considerations which prompted Lincoln to sign the bill also prompted the government to rush the project to completion. It was, in fact, completed in seven years, as a rush-order wartime measure. Here, again, one notes the familiar foreshortening of events, the telescoping of processes, which is so characteristic of nearly everything related to California.

The early completion of the Central Pacific, however, was related in still another way to the "peculiar" position of California. Most projects of this character had proceeded in an unilateral direction: from a base of operations to the fringe of settlement. But the Central Pacific was a two-way project. The rapid growth of California made it possible for Crocker and his associates to start two crews at the same time, one in the east, one in the west. The great difficulty with the western operation was the scarcity of labor. The Comstock Lode had been tapped by 1862 and Virginia City was a powerful magnet to men looking for work. Of one thousand men that Crocker recruited in California in 1863-64, only about two of every five reported for work; and, of those who did report, all but a few quit as soon as they had earned enough money to pay their fare to Virginia City. Faced with this emergency, Crocker, against the advice of his associates, began to import Chinese labor and, before the project was completed, some 15,000 coolies were at work. It would have been quite impossible to have completed the western end of the project on schedule had it not been for the use of Chinese labor and it was, of course, California's "peculiar" geographic position that made it possible.

THE PRECOCIOUS FRONTIER

For the first two decades after 1850 California was a state largely by virtue of an act of Congress rather than in point of social fact. "Politically," wrote Blake Ross, "California was a full-fledged state in the American Union, but economically and socially it was more like a colony characterized by frontier conditions." Its peculiar geographic position might have had entirely different consequences had it not been for the discovery of gold; but gold forced a many-sided development which took place in isolation. "History shows us by repeated instances," wrote Semple, "that the geographical conditions most favorable for the early development of a people are such as secure to it a certain amount of isolation." The California "frontier" was hardly like a frontier in the usual sense; it was a republic; not a colony. Because of the exceptional circumstances of its settlement and its remoteness from the centers of population, the California frontier was, as Franklin Walker has written, "a precocious frontier"; California "learned to talk while it was still young."

Its rate of growth alone set California apart from all American frontier communities. In 1848 San Francisco was a village of 800 inhabitants; twenty years later it was the capital of the Western empire, "the financial competitor of New York and a cultural rival of Boston." By 1860 California was producing its own food supply from 20,000 farms, and the value of its manufactures—astonishing for a "frontier" community—exceeded the value of gold production by twenty million dollars. The Fraser River "gold rush" of 1858, the discovery of the Comstock Lode in 1859, and the opening of the Arizona copper mines in 1870, had made San Francisco the mining capital of the West. Economically the states between the Rockies and the Sierra Nevadas were colonies of California instead of California being a "colony" of the East. The society which came into being in San Francisco with such magical swiftness was, as Walker has written, one which "had grown articulate enough in the days of its youth to speak while frontier conditions still existed." This was its charm and its uniqueness. Here in territorial isolation a society had sprung into being which was far more complex than the usual agrarian frontier.

In Whitman's phrase, the society which flourished in isolation in San Francisco for two decades was "fresh come, to a new world indeed, yet long prepared." In other frontier areas, the frontier experience was debilitating, wasteful of energies, leaving exhaustion in its wake. Here, in this frontier of silver-and-gold, the experience was exhilarating, tonic, exuberant. The other frontiers were debt-ridden and culturally impoverished; but not San Francisco. Thrown back upon their own resources for two decades, the Californians created, out of the wealth they possessed, a culture of their own. In the first decade of the gold rush, more books were published in San Francisco than were published in the rest of the United States west of the Mississippi; and, in the middle fifties, San Francisco could boast that it published more newspapers than London. Two daily papers were printed in French and others in German, Italian, Swedish, Spanish, and Chinese. Churches, theaters, and libraries appeared overnight. The per capita wealth of the state was the highest in the nation and the Californians used twice as much sugar and coffee, three times as much tea, and consumed "seven bottles of champagne to every one consumed by the Bostonians." It was not wealth alone which produced this astonishing social energy; but the coming-together, under unique circumstances and in isolation, of a score of national cultures, in a new and plastic society. It was not only a new but a very young society; as late as 1860, for example, two-thirds of the population was under fifty and it was, of course, predominantly a male society. "The dynamite of California," writes Walker, was "composed of one part vigor and one part unsatisfied passion." This, surely, was not the usual or typical frontier but "the acme of all frontiers, the most concentrated of quickly flourishing societies" in which "the people lived through *a condensed version of the world's economic and cultural growth*" (emphasis mine). Here, once more, note the familiar telescoping and foreshortening process.

Great significance attaches to the circumstance that, unlike other frontier societies, California learned to talk while it was still young. The uniqueness of its experience, in other words, was captured before the experience had ceased to be. The difference is that between an experience recaptured in memory, with all the distortions which time can work, and an experience caught up and immortalized while it was still being enacted. This difference has had the most important latter-day consequences. For it has given to the Californians a sharp, vivid, unforgettable image of this unique society, and this image has in turn influenced their subsequent social and political behavior. The gold rush image was not created in retrospect; it came into being simultaneously with the events and experiences which it reflected. The isolation of the California frontier forced those who created this image to look at what was taking place before their eyes, rather than to engage in sterile imitations of remembered experiences.

That so much of what was happening on this unique and precocious frontier found reflection in books, and stories, and memoirs, resulted in a legend of '49, of gold, of bonanzas, of great days. This legend is still very much alive and, by now, it has grown to enormous proportions, for it has been retold many times and continues to be retold. This legend is California's "past"; it is the unique cultural heritage of the people. The "past" means to the Californians not the Pilgrim fathers, or William Penn and the Indians, or George Washington crossing the Delaware; it means miners, and vigilantes, pan and rocker, the topsy-turvy of the gold camps, and San Francisco. The historically-minded Californian of today is orientated with reference to a set of meanings and significances quite unlike those by which the historically-minded in other areas, even in the West, guide their research and historical explorations. It is this circumstance which has made the collecting of Californiana a literary gold rush. The image of the "precocious" California frontier had, of course, far-reaching regional and even national cultural implications. Bret Harte's stories, as Van Wyck Brooks has said, were the prototypes of all the "Westerns" with all their stock characters that appeared in the later tales, just as his lynchings, holdups, stage robberies, and monte games were the models that hundreds of writers followed.

Union—With a Difference

On May 10, 1869, the last spike was driven in the Central Pacific, and San Francisco celebrated California's *real* admission to the Union—nineteen years after the act of formal admission. Prior to this time, the Californians had never thought of themselves as being part of the Union. Nothing expresses this relationship quite so clearly as the then popular phrase "back in the states." The "states" were always somehow external to California, far to the rear, almost as remote as Europe. This feeling was embodied in an editorial which appeared in the *Alta* on May 10, 1869: "California was formally admitted by an act of Congress to the sisterhood of the states nineteen years ago *but* that relation did not become a real, visible, tangible fact till the last rail was laid, and the last spike driven in the great continental road" (emphasis mine). This and other editorial comments suggest that the union was now consummated. "The states of the Pacific," as one editorial writer observed, "will no longer be divorced from the sympathies and affections of the *old* states" (again the emphasis is mine). Still another editorial, on this occasion, said that "the provincialisms which have grown up in our long isolation are doomed to a speedy death." Historians, surveying the period in retrospect, have echoed the same belief. "The beginning of a new era," writes Franklin Walker, "is always the end of an old one."

But did the "precocious" frontier belong after 1869 to the irrecoverable past? Actually the separation of nineteen years had worked a difference which did not vanish, and has not vanished. For, during these two decades, many things had happened in California: institutions had been planted; a culture had been evolved; and dozens of inter-related and highly exceptional cirumstances had combined to produce something new and distinctive. It is true that these "differences" tended to become less conspicuous after 1869, and that many of them were wholly erased or greatly minimized; but the sense of a subtly differentiated destiny could not be blotted out. Fifty years later, in visiting California, William James felt that he could sense here the promise of "the new society at last, proportionate to nature." For the factors which made, and still make, California "different" are not solely historical in character. The uniqueness of California's position is just as real today, in many respects, as it was in 1848. The intrusive alien influences which invaded the State after 1869 made serious inroads on the Californians' sense of an exceptional destiny; but the feeling was never completely obliterated. This sense of an exceptional destiny has always been rooted, as Blake Ross has written, in the belief that here, in this new land, the Californians were participating in a life quite different from that lived elsewhere in the United States. Those who had lived through the "days of '49" felt this difference in their bones; but one can still witness the exceptionalism of California at work in the remarkable transforming process by which, in the space of a few years, the children of Okie migrants become Californians to the manner born.

To understand the spirit of California, one really needs a sociology of what is called "good luck." An area that is poorly located, weak in resources, and handicapped by adverse historical circumstances eventually ceases to believe in its "luck"; in fact it comes to believe in its "bad luck" and this belief becomes a severe limitation on its development. In areas which have experienced a long period of depression, as in certain coal mining communities in Great Britain, psychiatrists have pointed out that the communities so affected become, in a very literal sense, "sick communities." The dead-end character of experience, the lack of incentives, the frustration that has beset successive generations, make for high rates of individual sickness, much of which is of a psychosomatic character, and also for a kind of community social sickness.

On the other hand, an exceptionally fortunate area comes to believe in its "good luck" and this belief becomes a positive, independent factor in the preservation of its good fortune. As subsequent chapters will point out, Californians have traditionally been reckless and self-confident gamblers; they have never hesitated to make high wagers against heavy odds and, on more than one occasion, have staked the future of the state on a throw of the dice, a turn of the cards. Many of these wagers have paid fantastic dividends and, on the average, most of them have been won. It is not that Californians are by nature inveterate gamblers or that people living in other areas are inhibited by a sense of inherited caution; the Californian gambles because he has confidence and he has confidence because his wagers have generally paid off. Although he is inclined, as all gamblers are inclined, to attribute his success to his "good luck," he also has an unformed, inarticulate awareness that the exceptional character of California is a sufficient hedge against almost any wager. He is like one of Bret Harte's gamblers whose insouciance conceals an awareness that the dice are loaded in his favor. Californians have always had this sense of being "lucky."

[5]

MIXED MULTITUDES

◇◇

CALIFORNIANS ARE not a unique people, but the "population" of California is quite unique. On the face of it, this would
seem to be an utterly inconsistent statement; but, properly understood, it goes far toward explaining the exceptionalism of California.
Populations differ in many significant respects: in their component
elements; in the way these elements are held together; and in
the manner in which these elements are interrelated or juxtaposed.
Composition, structure, and arrangement, however, fail to exhaust
the list of variables. Populations also differ in both a time and
space dimension. For example, is the population new to the area
or indigenous? Did the people come from short or long distances?
Lastly, all populations are changing or dynamic, but the rate of
change often shows marked variations.

In each of these dimensions, the population of California differs
from that of the other states. To such a degree is this true, that one
can fairly state that the exceptional character of the population
is one of the master keys to the contradictions and paradoxes of
the state. Yet, at the same time, the Californians are representative
of the American people. They are more like the Americans than
the Americans themselves—a riddle which this chapter will attempt to solve. The general frame of reference for an understanding of California's exceptional population is to be found in Davis
McEntire's suggestion that "California has come to represent to
the rest of the country what the United States has meant to the
world," namely, a land of exceptional opportunities. Here, in this
"great bowl of the west," the settlement of America has been
repeated but with a difference, a special accent, and, above all,

with a remarkable foreshortening of the process. California is all American but uniquely combined, uniquely put together.

GOLD SETS THE PATTERN

> Most of the Western states have been peopled by a steady influx of settlers from two or three older states. . . . But California was settled by a sudden rush of adventurers from all parts of the world. This mixed multitude, bringing with it a variety of manners, customs and ideas, formed a society more mobile and unstable, less governed by fixed beliefs and principles. . . .
>
> —LORD BRYCE

In studying the population of California, as in studying so many phases of the state's social history, the gold rush period provides, in microcosmic form, a preview of things to come. In manifold ways California has been reenacting the drama of the gold rush —at different levels, in different forms, but always in striking conformity with the underlying pattern set in the crucial first two decades. The gold rush changed the character of the westward movement of population and, in doing so, it set the pattern which California has consistently followed throughout the years.

The gold rush to California provided the first occasion for a general mixing-up of the diverse elements which had combined to form the new nation of the 1840's. Here, for the first time, the elements making up America came together and intermingled under circumstances which were not only unique but which, by their very nature, made for strange combinations and weird alignments. Until the gold rush, as James Truslow Adams has pointed out, "every American frontier had been settled by agriculturalists after the first advance of hunters and trappers and Indian traders. Except for the broad distinction between North and South, slave and free, plantation and farm economies, there had been a marked uniformity of social and intellectual life on all of them. *In this respect* the settlement of California offered a complete contrast. Every type of citizen of every social grade or profession came, not to hew forests, farms, and make homes, but to get rich as quickly as possible by a happy stroke of luck. Clerks, sailors, lawyers, doctors, farmers, even clergymen, anyone who loved adventure or believed in luck, tramped, rode or sailed to the newest promise in the Land of Promises." So universal was the interest in California, that the gold rush affected, as one historian has said, nearly every family in the nation.

In California, for the first time, Americans of every type, profession, background, and social class were thrown together in the great and compulsive confraternity of gold miners. There was a remarkable unity of purpose about this migration. Emigrants came to California, not to follow old and familiar pursuits, but to mine for gold. Twenty years after the gold rush, as one observer noted, you could find in any chance meeting of a dozen San Franciscans some one who had "worked in the mines." It was not merely the diversity of the elements which participated in the gold rush which was exceptional but, even more important perhaps, were the circumstances which precipitated the rush and which prevailed for two decades in California. Had these various elements been brought together under different circumstances, the mixing process might have produced quite different results.

Gold was the key to every aspect of the movement of people to California, a movement which has been properly characterized as one of the most extraordinary mass movements of population in the history of the western world. The attraction of gold accounts for the extraordinary diversity of types; the volume of the movement; and, above all, for its amazing velocity. Every emigrant in '49 was in an extraordinary *hurry* to reach the gold fields. The volume of this migration, in other words, must be multiplied by its velocity, to understand the unique combinations which it produced. Not only were the emigrants in a great hurry to reach the gold fields but, once there, the same energy kept them in motion, jostling them about, and sweeping them here and there. The various elements of this mass were constantly separating, coming together again, separating, and then recombining. There was no settling-down process. The mass was never stabilized. For the last hundred years, rapid population growth has been the normal, not the exceptional, characteristic of California's population expansion.

Gold also accounts for the fact that, unlike most population movements, the rush to California had an inverse order of progression. Those who were first to arrive on the scene had traveled the greatest distances. California was one of the truly remote sections of the world in 1848 and, since it was both a sea-and-land frontier, it was more easily reached by long sea routes than by shorter overland trails. The distance that people traveled had, of course, the most important social consequences. For one thing, it emphasized the selective forces at work in all migration movements. By and large, it was the younger, the more energetic and adventuresome elements that first struck out for the gold fields. Few women, children, or old people participated in the great pilgrimage to California. In 1850, 73 per cent of the population was concentrated in the 20-to-40 age bracket; 92 per cent were males; and the number of children was negligible. The further one is from "home," the more tenuous the home ties become, and California, by reason of its geographical position, was remote from all the centers of population, north and south, east and west.

Before proceeding to an analysis of the present-day population, it is worth while to pause a moment and to reflect upon the character of the gold rush migration. From 15,000 residents in 1848, the population of California soared to 165,000 in 1850, and, by 1860, stood at 379,994. From a thousand residents in 1848, the population of San Francisco jumped to 35,000 in 1851. Even these figures fail to provide an adequate measure of the volume and velocity of the gold rush migration, since thousands of emigrants came to California for a brief time and then left. Instead of slowing down, however, the tempo of population growth steadily accelerated: from 1848 to 1943 the average annual addition to California's population has been approximately 83,000, which is roughly equivalent to the number who came in the big gold rush year of 1849. These gold rush emigrants, it will be noted, had traveled great distances. Some had traveled 3,000 miles overland and others had come around the Horn, a route four times as distant. The recency of the migration is, also, worthy of emphasis. Hittell fixed the population of California at 340,000 in 1860 and, of this total, said that 325,000 were "immigrants or miners." Not one man in twenty was a native of the state in 1860 and, in 1948,

two out of every three people in California were born in some
other locality. Every state, nation, and race was represented in
the gold rush migration. The flags which could be seen flying in
the harbor of San Francisco in the late fall of 1849 told the
story: England, France, Spain, Portugal, Italy, Hamburg, Bremen,
Belgium, New Granada, Holland, Sweden, Oldenburg, Chile,
Peru, Russia, Mexico, Ecuador, Hanover, Norway, Hawaii, and
Tahiti. The names of the gold camps also indicate the diversity
of the migration: Washington, Boston, New York-of-the-Pacific,
Baltimore, Concord, Bunker Hill, Irish Creek, Italian Bar, French
Corral, German Bar, Dutch Flat, Kanaka Bar, Malay Camp,
Chinese Camp, Nigger Hill, Missouri Bar, Iowa Hill, Wisconsin
Hill, Illinoistown, Michigan Bluffs, Tennessee Creek, Kentucky
Flat, Minnesota Flat, Cape Cod Bar, Vermont Bar, Georgia Slide,
Alabama Bar, Dixie Valley, and Mississippi Bar.

But the pattern goes further and it has prevailed through the
years. This curious succession of groups tended to settle in
"islands" or "pockets." The gold rush migration was worldwide
in origin; nearly every group was represented but the groups
came at different times, in unique combinations, and in most uneven volume. Unlike most frontiers, California contained a large
foreign-born element: one-fourth of the population in 1850 was
foreign-born. During the gold rush decade, the largest foreign-
born groups were: Mexicans, English, Germans, Irish, and French.
No Chinese were reported in the census of 1850, but the census
of 1860 reported 35,000; that of 1870, 48,826. By 1860 the
Chinese were the largest foreign-born group in California; in fact,
every tenth person in California in 1860 was Chinese! Mexicans
were the largest foreign-born group in 1850, making up a third
of the foreign-born total; then, for some decades, they steadily declined in numerical importance only to regain, years later, the top
position. Since the gold rush coincided, roughly, with the Potato
Famine in Ireland, the Irish ranked fourth among the foreign-
born groups in 1850, and rose to first rank in 1870, when there
were 50,000 in the state (one-fourth of the foreign-born), and
then declined relative to other groups. This unique combination,
or juxtaposition, of cultural groups in California can, perhaps,

best be illustrated by the statement that, for thirty years the Irish and the Chinese competed for top rank among the foreign-born groups. In fact, the clash between the Irish and the Chinese forms an interesting chapter in the social history of the state. Where but in California have the Irish competed with the Chinese?

Germans were the third largest foreign-born group in 1850. This was due to the significance of the year 1848 in the history of the German people. From third place in 1850, the Germans rose to first rank in 1900, when the Germans totaled 72,000 (nearly a fifth of the foreign born), declined in importance for some years, and then jumped back to fifth place in 1940, when the number of Germans was approximately the same as in 1900. As can be noted from this brief statement of the constant expansion and contraction in the size of the foreign-born groups, the population of California is unique, not only in its diversity, but in the peculiar mode of succession, and the constantly changing relationships and combinations between the various foreign-born groups. About the only cultural minority which has shown little fluctuation in size is the Indian minority, which, today, is approximately the same size as it was in 1850.

The same pattern-of-succession appears in the "states-of-birth" groups, that is, the native born of other states. Most of the forty-niners came from the eastern seaboard states, those farthest removed from California, largely from New York, New England, and Pennsylvania. The second largest group came from Ohio, Indiana, and Illinois. The outstanding exception to this pattern was Missouri, which claimed 6,000 of the '49 migrants. These, of course, were the "Pike County" folk, celebrated in story and verse. Starting in the east, the point of origin for most of the migrants gradually shifted westward just as the point of origin for the foreign-born migration shifted within an ever-narrowing circle. The fact that migrants from other states came to California in this wave-like pattern, with one group succeeding another, provided a constant population dynamic. In the gold rush period, as today, the various cultural groups tended to be unevenly distributed in the state. For example, the Irish were concentrated in San Francisco; the Mexicans in Southern California; eastern seaboard migrants tended to settle in San Francisco; the Missourians in rural areas. In some cases, particular rural areas attracted migrants from particular states, as witness the concentration of Texans in Tulare County. It was not only the diversity of the gold rush migration, but the strange combinations that it produced both in spatial distribution and in time-sequence, that made it such a unique population movement.

If one correlates certain comments which J. S. Hittell made of California's population in 1860, it can be seen that the gold rush set the pattern for all the latter-day movements of population to California. Hittell noted, for example, that certain selective forces were at work in the gold rush migration. By and large, the immigrants were well educated for the time. It was rare, he said, "to find a white man who cannot read" and illiteracy was by no means exceptional in the America of 1860. Hittell found that the forty-niners had essentially the same traits as other Americans of the period but, *"the traits are more striking,"* because the immigrants made up, not a cross-section, but a *selection* of the American people. He was impressed, also, by the cosmopolitan character of the population and the astonishing social mobility that seemed to prevail. "High wages, migratory habits, and bachelor life" had made for a remarkable freedom from social restraints and taboos. "In no part of the United States," he observed, "is so much of life public, and so little private." Women were accorded a marked consideration and special deference. The mode of living tended to be luxurious; the tempo of social life was accelerated; and dancing, gambling, horse-racing, and billiard playing were "universal." Most of the migrants were young, energetic, and well educated by comparison with the norm for the period.

Make some allowances, here and there, for changes which have occurred, and Hittell's observations of 1860 have a remarkable pertinency today. In fact, the uniqueness of the population of California consists precisely in the fact that the striking characteristics of the gold rush migration have become more striking with the passage of time. If one projects the characteristics of the gold rush migration, and accelerates the tempo of change, it is possible to reach, by purely *a priori* reasoning, a workable understanding of the present-day population of California. Instead of settling down after the gold rush, California tended to become more diverse; the rate of growth accelerated; the unique patterning became more pronounced. This, of course, indicates the existence of a set of underlying dynamics which has given a remarkably consistent pattern to the expansion of California. The populations of other states, at one time or another, have shown some of the striking characteristics which have prevailed in California; but, in California, *the exception has been the norm* from 1848 to the present time. This becomes quite clear as one examines, under various heads, the more striking aspects of the population of California.

LIKE A GOURD IN THE NIGHT

Although the rate of population growth for the gold rush decade was phenomenal—the population quadrupled between 1850 and 1860—the real expansion of California dates from 1900. From 1900 to 1930 the population of the state increased from 1,500,000 to 6,000,000, an increase unprecedented in any other American state. In the decade from 1920 to 1930—another gold rush decade—the population increased 2,250,000, dwarfing the expansion for any prior decade. In the period from 1940 to 1948, new residents entered California at the rate of 270,000 per year, an average annual increase greater than that of the twenties and *three times greater* than the annual increase during the gold rush decade. These figures show that California has grown, as Lord Bryce said, "like a gourd in the night." A state which has grown at this clip, over a period of a hundred years, begins to take on special characteristics which are to be explained only in terms of the volume and velocity of its growth. Or, to put it another way, one can say that since the rate of growth in California has been consistently "abnormal," the population structure has acquired strikingly different characteristics.

A very high proportion of the residents of California from 1848 to 1948 have been born in some other locality. This is, indeed, as Marion Clawson has pointed out, "an overwhelming fact." It is not a fact which was once true and then ceased to be true; it has been a constant fact in the social history of California, a "normal" aspect of its development. It is as though, at regular intervals over a period of a hundred years, a state has been reborn, reconstituted, repopulated. Such a state cannot age, socially and culturally, except in a limited and relative sense; for it is kept "young" by periodic injections of new blood, of fresh energies.

To appreciate the impact of migration on the social structure of California, it should be pointed out that as the stream of migration has shifted from north to south within the state it has increased in volume and velocity. In 1848 southern California was older in point of settlement and more heavily populated than northern California. But, after the gold rush, the population was highly concentrated in the northern counties. For thirty years, southern California languished in isolation; even the completion of the transcontinental rail line in 1869 had, at first, little effect in the south. But, when the Southern Pacific and the Santa Fe finally extended their lines to Los Angeles, a spectacular land "boom" brought the first tidal wave of settlers.

By this time, of course, there were two "states" in California: Northern California, with its "old" and "settled" life and Southern California, with its "old" Mexican life, its gardens, orchards, and vineyards. As the stream of migration was diverted from the north to the south, the southern part of the state began to expand in the most remarkable manner. From one-fifth of the state's population in 1900, the southern counties acquired two-fifths by 1920 and by 1940 they had over half the population of the state. Not only was this migration thirty years later in point of time, but it had different origins and special characteristics, and was greater, both in volume and velocity, than the earlier migration to the northern counties. From seven per cent of the state's population in 1880, Southern California's proportion increased to 53 per cent in 1943. In the period from 1900 to 1940, the population of Los Angeles increased 1535.7 per cent by comparison with an increase of 172 per cent, for the same period, in San Francisco. The impact of migration, therefore, has been much greater in the south than in the north.

The sudden upsurge of southern California upset the balance of trade and industry within the state, brought about the most important political repercussions, and served to give new meaning to the geographical, social, and historical cleavage between the north and south. True to its generally topsy-turvy character, the "oldest" part of California is thus the "newest." From 1890 to 1930, the native sons have never constituted more than 27 per cent of the population of Los Angeles but they made up 66 per cent of the population of San Francisco in 1910 and 44 per cent in 1920. For many years, the conflict between "native sons" and "newcomers," which reflected the basic cleavage in the state, was an important factor in the social and political life of California. Being a seaport, San Francisco naturally attracted, at the outset, a larger foreign-born element than Los Angeles. Today one-fifth of the population of San Francisco is foreign-born by comparison with Long Beach, in Southern California, where the population is 90 per cent "white" and native-born. Not only has California grown faster than any other state, over a longer period of time, but portions of the state have grown much faster than other parts and have expanded at different periods of time. One might describe the effects produced by this differential rate of increase by saying that California has grown by a series of "explosions" occurring at different times in different parts of the state.

WHAT WAS YOUR NAME IN THE STATES?

> Oh, what was your name in the States?
> Was it Thompson, or Johnson, or Bates?
> Did you murder your wife
> And fly for your life?
> Say, what was your name in the States?
> —CALIFORNIA SONGSTER

Reflecting the westward movement of population in America, the point of origin for out-of-state migrants to California has gradually shifted westward. By 1890 the eastern seaboard states had given way to the east-north-central states as the principal source of migration to California. This wave of middle western migration, of course, was largely directed toward southern California. Migration from the inter-mountain western states did not reach significant proportions until about 1920. Then, in the 1930's, California began to attract a heavy migration from such states as Texas, Arkansas, Louisiana, and Oklahoma (migrants from these states made up about one-fourth of the total migration from 1935 to 1940). Although persons born in California outnumbered those born in any other state by 1860, they have constituted a steadily diminishing proportion of the population since 1900. Migration from Oregon and Washington to California has been important since 1860, but, in effect, this migration has represented a reshifting of population within the Pacific Coast region, since California has sent these states two migrants for every three which they have sent to California.

Not only is the population of California made up of representatives from every state in the union, but these states are represented in rough approximation to their numerical importance. These out-of-state migrants, moreover, have distributed themselves within California in a manner that reflects, in broad outline, the urban-rural balance of the states from which they came. Urban migrants have settled in urban areas; rural migrants in rural areas. Thus the cities of California are amalgams of the other cities of the nation as the rural areas are amalgams of rural America. In the San Joaquin Valley today one can find farm families from every important farming section of the United States. If the population of California had been consciously planned to make it representative of all America, its cities and farms, the results would not be much different from those which uncontrolled migration has produced.

From 1848 to the present time, migration to California has been overwhelmingly a one-way movement. In the period from 1935 to 1940, only 5,000 people moved from California to Oklahoma but nearly 95,000 moved from Oklahoma to California. The native-born Californians, morever, are among the least mobile people of any state. From 1860 to 1930, the percentage of persons born in California who have continued to reside in California has never been less than 90 per cent. The tendency of the native-born

Californians to stay in California is in marked contrast with, say, the tendency of the native-born Iowans to migrate: in 1930 one-third of the persons born in Iowa were living in some other state.

There has always been in California, however, an astonishingly large volume of intra-state migration. Between 1935 and 1940, 800,000 people—nearly 12 per cent of the total population—moved from one county to another. The rate of internal mobility is highest among the rural non-farm population, and reflects, of course, the seasonal crop cycle in California. But, if it were possible to secure figures on the manner in which population moves about *within* a particular city, as, say, in Los Angeles, the rate of mobility would probably be unbelievable. In Los Angeles an address or a telephone number is good, on an average, for only three or four months, and the telephone company reports that thousands of families have as many as three and four telephone numbers in the course of a year. This milling-about is most noticeable in those areas which have received the heaviest migration, and in which the velocity of migration has been most pronounced. Migrants pour into many California communities with such speed that they are kept spinning about for months and often years before the momentum with which they arrived can be broken or checked.

A QUILT OF MANY COLORS

As might be expected, the foreign-born population of California is today more diverse than it was in 1850. In 1850, Mexicans, Germans, British-born, Irish, and French made up four-fifths of the foreign-born total; but in 1940, twelve groups, not five, made up the four-fifths category. The largest groups in 1940, in the order of their numerical importance, were: Mexicans, Italians, British, Canadians, Scandinavians, Russians, Irish, Japanese, Portuguese, French, Swiss, Chinese, Austrians, Poles, Greeks, Yugoslavs, immigrants from the Azores, Spanish, Dutch, "other Asians," Hungarians, Finns, French Canadians, and South Americans. To give some idea of this diversity, suffice it to say that in 1940, of the "white" population of California, there were about 1,500,000 residents who spoke some foreign language.

While the foreign-born element has grown more diverse in California, it has declined as a percentage of total population. In fact, the foreign-born total has declined as the out-of-state total has increased. The foreign-born made up 24 per cent of the population in 1850, 38 per cent in 1860, and 37 per cent in 1870. From 1860 to 1870, nearly 40 per cent of the population of the state was foreign-born. From 1890 to 1920, the foreign-born still made up nearly a third of the total migrants to California and, in the decade from 1920 to 1930, nearly 450,000 foreign-born persons settled in the state. However, the increase in the American native-born has been so great since 1920 that the foreign-born now constitute only 13.4 per cent of the state's total population.

The foreign-born colonies are spotted curiously on the map of California. Nearly half of the Irish, the French, and the Italians are in San Francisco, while 70 per cent of the Mexicans and a similar percentage of the Russians (mostly Russian Jews) are in Los Angeles. The Chinese are San Franciscans; the Japanese are Angelenos. Most of the Indians and the Negroes reside in the southern counties. In San Francisco in 1940 one "white" person in every five was foreign-born; in Los Angeles the ratio was one to seven. This uneven geographical distribution of the foreign-born has accentuated the diversity in types, the mixed-up character of the population.

In other parts of the country outside the South, the most mixed populations, racially and culturally, are commonly found in the large cities; and the farm populations are, by comparison, relatively homogeneous. Exactly *the opposite* pattern prevails in California, where the most heterogeneous population is to be found in the rural areas, on the farms. In 1940 more than one person of every three in the farm population of the state was either foreign-born or had foreign-born parents, and 10 per cent of the farm population was "non-white." On the other hand, the "small town" population in rural California is very largely made up of native-born "white" elements. In other states, the "farm people," the "downstate" or the "upstate" people, are invariably known as the "backbone" of the white, Protestant, Anglo-Saxon element; but, in California, a large percentage of the "farm folks" are Italians,

[75]

Portuguese, Japanese, Filipinos, and Scandinavians—Protestants, Catholics, and Buddhists. One might say that California's real melting pot is in the rural areas. This upside-down relationship accounts for many curious aspects in the social history of California. Thus California has a "farm vote," but it is a farm vote with a difference. There are great stretches of farming country in the east and middle west which have never known a clash of cultures or a conflict of races; but racial conflicts and cultural clashes have been known in rural areas in California since the earliest times.

In most areas of the United States, the racial pattern is two-dimensional—Negroes and whites—but in California it has four dimensions: Negroes, whites, orientals, and Mexicans. One could add a fifth dimension by the inclusion of Indians but, for many years, Indians have not been a major factor in the state's racial problem. This difference in the pattern of race relations has created a fairly favorable atmosphere for Negroes, since the Chinese, Japanese, Filipinos, and Mexicans have borne the brunt of the attack which has, from time to time, been leveled against racial minorities. The number of Negroes has increased in Los Angeles from 188 in 1880 to 165,000 in 1948. Thus today California has a sizable representation of every racial type to be found in the American population: Negroes, Indians, Orientals, Filipinos, Hindus, Mexicans (a mixed type), and whites. In no other states can all these elements be found, in significant numbers, living side by side. California is, indeed, a quilt of many colors.

OLD AND YOUNG IN CALIFORNIA

Since only a small part of California's population growth has been due to natural increase, the excess of births over deaths, the age structure of the state's population has always been somewhat "abnormal" or, more accurately, exceptional. The current folklore has it that the state is dominated by oldsters or "senior citizens," but the facts, which are admittedly complex, indicate that almost the reverse is true. From about 1880 to the present time there have been two major groupings in the state's population: those born in California and migrants. The population structure, usually shown in the form of a pyramid or "pine tree," assumes quite different forms when these two groups are studied separately. Since the young migrate more readily than the old, the migrants to California have been, on the whole, somewhat "younger" than the settled population of the state. Although it is true that many old people migrate to California, particularly to southern California, the older-age groups have always made up a relatively minor proportion of the total migration. Most of the migrants have been young, in the 20-to-50 age bracket, but they have not brought many children, 10 years of age or younger, to California. On the other hand, the California-born residents show a heavier concentration in the younger age groupings.

When the two groups are combined, however, it appears that California has a population deficit in the 1-to-25 age category (it has only about four-fifths as many people in this category as the nation as a whole); but that it has 10 per cent more people in the 25-to-29 category than the national average. Above the 50-year level, California has about 20 per cent more people in the older-age categories than does the nation as a whole. Although it is true, therefore, that the California population is somewhat "older" than the nation's, a paradoxical situation in a "new" state, this abnormality arises more from a deficit of youngsters than from a surplus of oldsters. What is more important, however, is the fact that in the age group from 25-to-50, which represents the most productive age category, California has a higher concentration than the national average. If this category is broken down still further, it develops that California has a 20 per cent surplus in the 25-to-30 category. The rising average age of the population in California follows a national trend but, as might be expected, the trend has gone further in California than in the country as a whole.

There appear to be more old people in California than the facts reveal largely because persons over sixty are unevenly distributed in the state. The "senior citizens," rich and poor alike, have shown a strong preference for certain communities, a circumstance which has given rise to the California population phenomenon of "retirement colonies." In such communities as Santa Cruz and Los Gatos, in northern California, and Pasadena and Sierra Madre, in southern California, persons over sixty make up a fifth of the total population, which is nearly twice the average for the state. On the other hand, the "over sixty" folks make up only 6 per cent of the population of the Imperial Valley towns which were founded after 1900, which is less than half the state average. The "over sixty" category has a special significance in California because of the concentration of older people in "retirement colonies," and because more older people live upon income from investments than would be true, for the same groups, in other areas.

The abnormal age distribution of the California population has had certain important consequences. Having fewer young children to educate than other states, California has been able to maintain enviably liberal provisions for its public schools. Few states have been more generous, in this respect. California, however, has a peculiar "old age" problem since the proportion of old people to working people is somewhat higher than in other states and will be still higher in the future. But perhaps the most important consequence of the abnormal population pyramid consists in the fact that California has a per capita income 40 per cent above that of the nation as a whole which reflects, of course, the concentration of population in the 20-to-50 category. In good times and bad, California's per capita income has been consistently higher than that for the nation. Marion Clawson estimates that at least one third of this differential is due to the peculiar age distribution of the population. A low rate of natural increase and a preponderance of "older" people have also made for a slower rate of growth in the labor force which in part accounts for the fact that California is an area of high wages. These population differentials, in short, are some of the "hidden" secrets of California's "good luck." If migration should decline in volume, per capita incomes would tend to fall in California.

To some extent, the population pyramid in California is now tending to approximate more closely that of the nation's. A very large percentage of the war-time migration was made up of young people. As a consequence, birth rates have shown a sharp increase. The birth rate has risen from 12.6 per cent in 1933 to 16.2 per cent in 1940 to 21.9 per cent in 1943. As the number of children in the younger-age group increases, the school system must be rapidly expanded, a problem that has already reached acute proportions. At the same time, California will have, for many years, a serious problem with its old-age, or pension groups, of which it has a relatively high percentage.

THE ORIGIN OF GALLANTRY

During the past century, California has been predominantly a masculine state with the usual tendencies toward gallantry and "equal rights" that such states exhibit. Since men migrate more readily than women, a population based largely on migration will naturally show a preponderance of males. In the gold rush decade, the ratio of males to females in California was 12 to 1. In 1860, women made up only 30 per cent of the population and, a decade later, 37 per cent. Although the ratio of the sexes has tended toward a balance, the 1940 census showed that men still outnumbered women in California in the ratio of 104 males to 100 females. In this respect, California follows a national trend but in a more striking manner, and with special emphasis.

J. S. Hittell, in a somewhat lugubrious vein, once said that the ratio between the sexes in California was "unsound." Sound or unsound, it has certainly been most inconvenient at times. Hittell was disturbed to find an exceptionally large number of women in the mining camps who were, as he said, "neither maidens, wives, nor widows." A barroom ballad, once popular in California, defined these ladies more accurately:

> The miners came in forty-nine,
> The whores in fifty-one;
> When they got together,
> They produced the native son.

Hittell was also keenly distressed by the low birth rate in California but, being an incorrigible boaster, he rationalized this advantage by saying that California-born children were "larger" at birth than children born in other states and that, in any case, more twins were born in California than elsewhere. He was also convinced that residence in California made for exceptional fertility. "It has been marked," he wrote, "that a multitude of instances have occurred of couples who, after having lived childless for ten, fifteen, or twenty years in other countries before coming to California, in a year after arrival here have had children. Traveling and a change of climate," he noted, "will no doubt always have a favorable influence in this respect, but perhaps the extraordinary productiveness of California may be perceptible here too."

Since California has always been, in some respects, "a mining camp," it is not surprising that the peculiarly harsh ratio of the sexes found in the camps had its counterpart in the state's agricultural labor camps. Most of the immigrants from the Orient have, of course, been men. As late as 1940, foreign-born Chinese men outnumbered foreign-born Chinese women in the ratio of 4 to 1; there were nearly 9 Filipino men for every Filipino woman; and, among the Japanese, there were 160 foreign-born men for every 100 foreign-born women. Among the foreign-born as a whole, the ratio has been 121 men for every 100 women. Waving these figures aside as utterly irrelevant, the Californians consistently berated the Chinese for their traffic in "slave girls" which was always grossly exaggerated); damned the Filipinos for their addiction to dancehalls; and complained bitterly about the "picture brides" imported from Japan.

Where women are at a premium, women's rights are freely granted. A visitor to California in 1867 wrote that "under social arrangements so abnormal, a white woman is treated everywhere on the Pacific slopes, not as man's equal and companion, but as a strange and costly creature freed from the restraints and penalties of ordinary law." This special consideration shown the ladies often amounted, it was said, to excessive indulgence. "The American women of California," wrote Hittell, "are not healthy. . . . They are trained up in dark and idleness, as though sunshine and work would ruin them. Pastry, pickles, and sweetmeats form a considerable portion of their food, and they are taught to abhor coarse strength and robustness as worse than sins."

The California gold rush had a marked effect upon the divorce rate in the eastern and middle western states as wives secured uncontested divorces from husbands who had left for the gold fields. Don C. Seitz has even suggested that the discovery of gold in California was responsible for the extraordinary number of spinsters in New England. In California divorces were naturally looked upon with favor and were freely granted. And, as Hittell observed, "when a woman is oppressed by her husband, in California, she can generally find somebody else who will not oppress her." The gold rush historians have pointed out that the divorce rate which prevailed in California was the highest in the world and that the plaintiffs were invariably women. The preponderance of men is unquestionably the key to the "easy" divorce laws of the western states, and it has a bearing on the consistently high divorce rates which have prevailed in California.

At the end of the nineteenth century, California was still essentially a man's state. It is precisely in such states, as Arthur W. Calhoun has pointed out, that "woman rises to sovereignty." California adopted, in its first constitution, the Spanish law of "community property," and took care to provide that the wife's property, both real and personal, which she had at the time of the marriage, should remain her separate property. Both provisions were most liberal by comparison with the law of marital relations in other states in 1849. Since the wife retained, as her separate property, whatever property she possessed at the time of marriage, and since she was given an equal share in all the husband's earnings after marriage, a remarkable independence was conferred upon women in California. Although women are no longer regarded as "strange and costly creatures" in California, the tradition prevails as shown by the fact that California has always been most progressive so far as women's rights are concerned. I hasten to add, however, that women are no longer "trained up in dark and idleness" nor are they nowadays addicted to "pastry, pickles, and sweetmeats."

A State of Cities

Unlike other frontier communities, California started off with a high level of urbanization; in fact, the level of urbanization was higher in California than in any other section of the country except the North Atlantic States. As early as 1870, California was among the ten most urban states in the country, and its great increase in population, through the years, has been primarily a growth in urban population. Today nearly three-fourths of the population of California lives in seven metropolitan districts. The Los Angeles and San Francisco-Oakland areas account for 62 per cent of the state's population, and 42 per cent live within the Los Angeles metropolitan area alone. Only three states, New York, Pennsylvania and Massachusetts, have a higher percentage of people living in cities than California. In every decade since 1850, California cities have grown more rapidly in population than the rural areas, even during periods of great agricultural expansion. Similarly, the proportion of the state's population living in rural areas has declined at every census from 1850 to the present time. To a very large extent, therefore, California's increase in population through migration has represented a shift from rural to urban residence as well as a westward shift in population. The remarkable urbanization of population in California reflects, of course, a national trend, but, here again, the trend has proceeded faster and farther than in most other states.

When one considers that California was, by every standard, a "frontier" state in 1870, it is, indeed, remarkable that it should even then have ranked among the ten most urban states in the country. The fact that it did suggests the existence of a peculiar relation between California and the other western states. This relation might be defined by the statement that California is the urban part of the West; that it functions, in other words, as the urban center for the other western states.

Still another key to the urbanization of California's population is to be found in the fact that California has always been able to derive a major share of its total income from agriculture, without increasing its agriculture population in relation to its urban population. This it has been able to do because of the peculiar character of its agriculture. One of the richest agricultural states in the Union, California had only ten per cent of its population living in rural areas in 1930. From 1860 to 1940, the urban population of the state has increased sixty-fold but the rural population has increased only seven-fold, despite a fantastic increase in the volume of agricultural production. If one thinks of populations as being divided into two main classifications—urban and rural—California's rural population is much smaller than one might expect in terms of its fabulously productive agriculture.

Within the "rural" classification, however, another discrepancy appears. Less than a third of California's "rural" population in 1940 lived on farms. The number of people living in rural areas, but not on farms, increased in the 1930's by 432,000, a growth of more than 46 per cent; but this was largely in the rural non-farm category. What this means, of course, is that even agriculture in California is highly "urbanized"; a large proportion of those engaged in agriculture live, not on farms, but in small towns and cities in the rural areas. This is to be explained, again, by the character of the state's agriculture, and also by the fact that as distributing centers, the cities and towns of the interior of California have always been at a disadvantage, in terms of freight rates, with the coastal cities. In the inception of a trend toward urbanization in California, and the constant acceleration of this trend, one can see that certain historical forces have also been at work. The fact that California got off to oo much more rapid a otart than the other western states, for example, is doubtless a factor in its ever-increasing urbanization.

POPULATION WHIRLIGIG

The states of the Ohio and upper Mississippi Valley have always been characterized by a high degree of homogeneity which has reflected, as Dr. Dan Elbert Clark has written, "a general similarity of physical environment and common purposes and needs," and, also, a similar pattern of settlement through the gradual extension of the frontier. Similar observations might be made of both the South and the West; but no such homogeneity has ever existed in California. For California has always occupied, in relation to the other regions, much the same relation that America has occupied toward Europe: it is the great catch-all, the vortex at the continent's end into which elements of America's diverse population have been drawn, whirled around, mixed up, and resorted. Basically, the explanation for California's diversity is to be found in three factors: the reasons which have prompted people to move to California; the diversity of resources in California; and its geographical position.

Since California was essentially a mining *and* agricultural frontier, it naturally attracted a different type of migration than that which was drawn to the usual frontier settlement. Mining frontiers always attract a great diversity of types. From 1848 to the present time, people have been attracted to California for a variety of reasons and the diversity of motivations has been reflected in the variety of types. Through the years, thousands of people have come to California primarily because they found the climate attractive; because California was a pleasant place in which to retire; and for a variety of non-economic motivations. The discovery of gold made California world-famous, and curiosity alone, has drawn thousands of migrants westward. Capitalizing on the gold rush legend, California has consistently sought to attract new residents through aggressive promotion campaigns often financed by grants of public funds. The non-economic attractions of California have not been limited to any one class, or type, or group, but have had the widest possible appeal. For better or worse, the legend of California as a "Land of Promise" is now too firmly rooted in the consciousness of the nation to be offset by "warnings," hostile legislation, or other measures aimed at diverting the flow of migration.

The reasons which have prompted people to migrate to California have also been closely related to the diversity of opportunities in California which in turn reflects the great geographical diversity of California as a state. The geographical diversity of California is so great as to make of it an anomaly even among states which also show a great range of environmental conditions. In this sense, California is beyond any doubt a special case. In the variation of climatic conditions; in the diversity of its soils; and in contrasting topography, California shows a truly astonishing diversity. "In no similar area in North America," writes Dr. R. T. Young, "are there such great extremes of climate or more marked differences in the corresponding life. . . . Even the *flora* and *fauna* of California are peculiar to themselves." There is, however, a special feature about the geographical diversity of California. "The conspicuous fact about California's environment," writes Dr. John Walton Caughey, "appears to be its versatility." California's environment is certainly one of the most versatile, the most plastic and adaptable, to be found anywhere in the world. The diversity of California's agriculture is almost matched by the diversity of its mineral production. Economic geographers have consistently emphasized that California is one of the most perfectly balanced, most nearly self-sufficient regions in the nation. It has been able to offer, therefore, a wide range of opportunities in almost every significant field of economic activity.

The geographic location of California has also been of prime importance in attracting a most diverse migration. It has been to Asia what New York has been to Europe: the first landing place for east-bound migrants from across the Pacific. It has also been the terminus of the westward movement. Chinese, Japanese, Filipinos, and Russians have come to California from the East. By reason of its position, it has also drawn migrants from Mexico, Central and South America. From many points of view, therefore, California has occupied a unique geographic position which has enabled it to attract migrants from all corners of the world. The completion of the Panama Canal, for example, greatly shortened the distance, measured in time, between California and Europe. The expansion of population in Europe has long blinded people to the central position which California occupies on the global maps. In the last century, California has steadily moved toward a more central position in world affairs. In 1848 it was one of the most remote areas of the world, the last frontier of America, in a world in which the United States still occupied a subordinate position to Europe. Today, the United States is the world's first power; and California, no longer a frontier, occupies a more central position, in terms of our global interest, than the older settlements on the eastern seaboard. The movement to locate the headquarters of the United Nations in San Francisco was a symbolic recognition of the fact that California, in geophysical terms, today occupies a central position in world affairs.

The marks of migration can be seen, not only in the diversity of

California's population, but in the curious manner in which the migrants have been re-grouped within the state. "Migration," wrote Dr. Robert E. Park, "has had a marked effect upon the social structure of California society . . . where a large part of the population, which comes from *diverse and distant* places, lives in more or less *closed communities,* in intimate economic dependence, but in more or less complete cultural independence of the world about them" (italics mine). Just as the aged have their "retirement colonies" in California, so one can find colonies of Portuguese dairy farmers, Armenian raisin-growers, Yugoslav fishing colonies, Japanese produce-farmers, as well as a miscellany of Chinatowns, Mexican "jim-towns," and Russian Molokan settlements. Even the rich have their colonies. Pasadena, for example, is the home of families whose wealth is based on inheritance; the rich European refugees are clustered along the Riviera; and the nouveau riche are to be found in Beverly Hills and Bel-Air. For the disposition of racial and ethnic minorities to settle in colonies finds its counterpart in the disposition of social classes to segregate themselves. Dr. Park's characterization of Southern California as "a congeries of culturally insulated communities" can be applied, with some modifications, to the entire state.

If California's population represented a thorough cross-section of the American people, its diversity might have less significance for the differences would tend to cancel out. But the population of the state really represents a selection, rather than a cross-section, of the national population, and a selection of this sort tends *to heighten,* to emphasize, the diverse traits and characteristics of the populations from which the migrants have been drawn. Throughout this chapter, I have had occasion, again and again, to stress the point that national population trends are heightened in California; that they appear in a more extreme form here than elsewhere. This tendency reflects the fact that the population of California has been selected, rather than drawn at random, from every state, every class, every race, every ethnic element in the American population. Immigration to America has, also, been selective; but a selection from an already selected population brings into sharp focus the more striking traits and characteristics of the base population.

[*86*]

It is for this reason that one can say that the Californians are more like the Americans than the Americans themselves. Lord Bryce, who saw California clearly and saw it whole, observed that the Californians were "impatient . . . for the slow approach of the millennium" and were always "ready to try instant, even if perilous, remedies for a present evil."

But there is still another dimension to the impact of migration in California which must be explored even if it cannot be exactly defined. If one can imagine a situation in which a *selection* of the American people had been drawn to a region the environment of which was more like that of the rest of the country than California's, the social and cultural consequences would be entirely different from what they have been in California. For California *is different*, and the totality of its differences has been brought to bear upon its selected population. If all of the migrants to California had come from a single place of origin, or if they all had a common cultural background, *and* if the physical environment into which they moved, had not been strongly different from the environment which they left, more of the customs and traditions of the area of origin would have been transplanted in California. But, as Marion Clawson has said, "the very diversity of California's migration, *as well as* the fact that the physical environment in the state differs from that in the areas of origin, has led to some abandonment of the old forms and traditions and to the evolution of certain new ones."

Migration naturally tends to weaken the well-established family ties, social customs, and the traditions of migrants; but, in California, this tendency has been given a special emphasis by reason of the novelty of the environment. In surveying the culture of California, one will find in every field—in mining, agriculture, industry, technology, gardening, architecture—that the novelty of the environment and its compulsive quality have *forced* an abandonment of the imported cultural pattern in many important respects. "So great a departure from the climate of the Midwest and East," reads an early report of the Commission of Agriculture, "subjected the culture of the soil to novel conditions, unsettling old traditions, and defying some of the most tenaciously held lessons of experiences in the older parts of the country." From a statement of this sort, one turns to a recently published work, *Pacific Coast Gardening,* by Norvell Gillespie, and is there informed that both the technique and timing of gardening on the Pacific Coast are quite unique and that success depends on a know-how which cannot be imported but must be discovered here, by a painful trial-and-error process.

The culture of California has two striking characteristics: the willingness of the people to abandon the old ways, *and* the willingness with which people will try new forms and modes, and the inventiveness which they show in devising such modes and forms. These inter-related dynamics are at work in every phase of the culture of the state; in its politics no less than in its agriculture. Migration accounts for the weakening of old ties but it is the challenge of a novel and highly versatile environment which explains the inventiveness, the quickness with which something new is devised. This latter statement, however, requires some refinement. The inventiveness of the people can be measured by the degree to which the new environment differs from the old. Kansas differed from Illinois as Illinois differed from Pennsylvania, but the degree of these differences is slight when measured against the degree by which California differs from all of these states. In discussing the diffusion of cultures, Arnold Toynbee has suggested that: The greater the difficulty of the environment, the greater the stimulus to inventiveness; and the related generalization that new ground provides a greater stimulus to activity than old ground. California was both new and difficult. Its difficulties consisted not in a meagerness of resources but in the fact that its resources could be unlocked only by untried, freshly devised methods. In these dynamics is to be found the basis for the belief, dating from the first American impressions of California, that this state held, as William James phrased it, the promise of "the new society at last, proportionate to nature." [1]

[1] For an interesting account of the relationship between mobility and crime, see: Final Report on the Special Crime Study, Commission on Social and Economic Causes of Crime and Delinquency, Sacramento, June 20, 1949.

[6]

CALIFORNIA LATIFUNDIA

◇◇

ON MARCH 14, 1850, a hundred or more armed settlers, most of whom had been in California only a few months, marched through the streets of Sacramento, and, at a mass meeting, announced that henceforth they would oppose with force and violence all attempts by the courts to eject them from lands which they had occupied. Replying to this demonstration, the landowners summoned a *posse* and, as two groups collided in the streets, a riot ensued in which three men were killed, including the city assessor, and many more were injured. The leader of this squatters' riot was Dr. Charles Robinson, of Fitchburg, Massachusetts, who, a few months previously, had "crossed the plains" to California. After recovering from wounds received in the riot, Dr. Robinson was elected to the state legislature. (He later became governor of Kansas.) To restore "law and order," the state militia was summoned to Sacramento. For many years after the riot, the Settlers' or Squatters' Party formed an influential faction in state politics. How then, did it happen that in "frontier" California, on the eve of the state's admission to the Union, land-hunger could have reached a point where it precipitated an armed insurrection? Wasn't there land enough and for all in the California of 1850? What was the background of this anomalous incident?

FRONTIER WITHOUT HOMESTEADS

Few factors had more to do with the rapid westward expansion of the American people than the "free lands" policy of the federal government. To a greater or lesser extent, every state west of the Alleghenies experienced the leavening effect of the free or

[*89*]

"cheap" land policy of the federal government in the administration of the public domain; that is, every state except California.

When the tide of western migration began to pour into California, the settlers discovered, with anger and amazement, that a large area of the best land had already passed into private ownership. By the time of the American conquest, in fact, most of the arable lands of the state were held by a small number of large land-owners who had obtained their holdings by grants from the Spanish Crown or the Mexican government. Although the census of 1850 reported only 872 farms in California, the average size of these holdings was 4,465 acres. Following the secularization laws of 1833, which broke up the Mission estates, nearly 800 private land grants, embracing about 26,000,000 acres or nearly one-fourth of the lands of California, had been made by the Mexican government. After investigation by an American commission, 588 of these grants, totaling 8,850,143 acres, or an average of 15,051 acres each, were confirmed. This vast area, needless to say, was automatically removed from the operation of the Pre-emption Act of 1841 and the Homestead Act of 1862.

To appreciate the extent to which lands had passed into private ownership in California prior to 1848, it is important to remember that large areas of the state were, and still are, unfit for farming. Of 101,563,520 acres making up the total land area of California, 12,895,000 acres were listed in 1940 as "land available for crops" and 6,831,000 acres as "land used for crops." Actually the core of the state's agriculture consists of about 3,500,000 acres of irrigated cropland; the return from these lands makes up about 83 per cent of the total agricultural income. The importance of the area embraced in land grants, an area that was never open for free home-stead settlement, is, therefore, much greater than the actual acreage figures indicate. For this area included some of the state's finest agricultural lands and the good land was limited in California. When one realizes that about one-fourth of the land area of the state was embraced in land grants (although the acreage actually confirmed was greatly reduced) and that the remaining three-fourths was agriculturally of little value, then the importance of the pre-statehood pattern of land ownership can be properly appreciated. Whatever effect the Homestead Act may have had upon the development of agricultural patterns in other sections of the country, the settlement of California proceeded with only slight conformity to the theory and intention of this act.

Despite the largesse with which lands had been granted to private owners prior to 1848, the pre-American pattern of land tenure could hardly be regarded as monopolistic in terms of the uses which then prevailed. For the pre-American grants had been made, as J. S. Hittell pointed out, "to suit the habits and wants of the people" and there were probably not more than 10,000 people in California in 1846. The native Californians owned large herds of cattle but these cattle were not fed on cultivated food or kept in fields or placed under shelter. Since a nearly unbroken drought prevailed between May and November of each year, it took an enormous amount of land to support a herd. The land grants, therefore, were not nearly as "large" as they appeared to be to those who were unfamiliar with the uses to which the land was put. Under Spanish and Mexican rule, the typical land unit was the square league or about 4,438 acres; but this unit could not be compared with an economical land unit in, say, the Ohio Valley. Actually the California grants were not much "larger" in terms of need than the peculiar semi-arid conditions which prevailed in the state warranted. "By nature and tradition," writes Dr. Frank Adams, "the Californians were ranchers rather than farmers."

At the outset, therefore, a sharp conflict developed between settlers, who were imbued with the Anglo-American concept of a "farm," and the owners of the large land grants and their successors. The settlers came to California with the belief that, except for a few settlements along the coast, all the land in the territory was public domain and therefore open to settlement. One can imagine, therefore, their disappointment when they discovered that thousands of acres of the best land of the state, lying uncultivated, were claimed by a small number of landowners under some inchoate loose grant made by the hated Mexican government. To appreciate this feeling of disappointment and resentment, however, one must realize that these settlers had traveled great distances to reach California and had endured untold hardships en route. Needless to say, it was just as difficult to leave California, in 1848, as it was to reach it; once there settlers had no easy means of escape. In a memorial address to Congress in 1850, the settlers called attention to the hardships they had endured in "crossing the plains" and re-affirmed their belief that California belonged to the United States and hence was open to settlement. On arriving in California, however, they had discovered that the land was monopolized and that they were, in legal effect, trespassers. Out of this bitterness and frustration, came the squatter riots of 1850. Basically the early land conflict in California involved a conflict between two traditions: the Anglo-American farm tradition and the Spanish *hacienda* or rancho tradition.

This conflict, at the outset, centered in the dispute over land titles, a dispute which raged in California from 1850 to 1870. Under the Treaty of Guadalupe Hidalgo, the United States had agreed that property of every kind belonging to Mexicans in the ceded territories should be "inviolably respected." Although the guarantee was clear-cut and specific, it so happened that many of the grants were subject to attack: for technical non-compliance with the laws of Mexico; for vagueness and uncertainty in the designation of boundaries; and, also, because of the loss of validating documents. Some of the grants were tainted with fraud and others were outright forgeries. The public lands had not been surveyed prior to 1848 so that there were no basic maps or documents to which grants could be related. In view of these circumstances, the American Claims Commission, established in 1851 to pass on the grants, adopted a policy that was quite liberal to the claimants and most of the grants were validated. Once the grants were validated, of course, the lands could be subdivided but, during the long period of confusion and uncertainty which prevailed, settlement of the land was retarded. The uncertainty about titles, therefore, was possibly as much a factor in land monopoly as the size of the grants. It has been estimated that this uncertainty about titles and the delay in their validation, prevented the settlement of from five to twenty times the quantity of land which the terms of the grants called for.

The real beneficiaries of this twenty-year period of confusion in land titles were the land speculators who, without exception, were American claimants, by assignment, of Mexican grants. The principal victims, of course, were the American settlers or squatters and the original Mexican claimants. The Mexican claimants lost out early in the struggle since they lacked the resources required to take a case all the way from the Claims Commission to the Supreme Court of the United States. Had it not been for this twenty-year period of confusion, strife, and litigation, the large grants would, in the normal course of events, have been broken up and subdivided. But this breaking up process was abnormally deferred. The net result of the conflict over land titles was that the promise of the American frontier was never fulfilled in California. "During the first quarter century of American occupation," writes Dr. Adams, "the confusion in land titles so discouraged settlement that thousands who had come to California to acquire farms returned to their former homes in disgust, and other thousands, learning of the difficulties, stayed away. Most of the grants gradually passed into the hands of Americans, some to continue as great cattle ranches, others to be held for speculation that further retarded subdivision and settlement."

For a long time now it has been the fashion in California to minimize the latter-day importance of the Spanish and Mexican land grants and to emphasize merely the "romance of the ranchos." But recent investigation has shown that many of the present-day large farms and ranch properties of California represent residual portions of original Spanish or Mexican grants which were never completely broken up or subdivided. On the basis of information provided by county assessors, Dr. Adams has shown that 19 of the original grants, all in grazing sections, are still intact; that at least 12 additional grants are still assessed to single ownership to the extent of two-thirds of their original area; and that there are 6 additional grants in single ownership which cover from one-third to one-half of the area embraced in the original grant. Although the old grant lines have been obliterated, in most cases, by subsequent subdivision and resale, in some of the areas formerly in grants "the patchwork shape of individual parcels and the location of roads and persistence of fences along grant lines have resulted in an agricultural landscape that contrasts strikingly with the checkerboard pattern of areas divided on the basis of public land surveys." The reason for this variation in the pattern, of course, is that California largely skipped the frontier phase in the settlement of its agricultural lands.

THE RAPE OF THE PUBLIC DOMAIN

As I have indicated, 588 claims, based on private land grants and totalling 8,850,143 acres, were eventually confirmed in California; but this still left a vast acreage that was theoretically open to settlement. What happened to this additional acreage? For one thing, the "boom times" which the discovery of gold ushered into being in California, created an unprecedented opportunity for land speculation. From 1862 to 1880 public land sales and warrant and scrip entries in California surpassed similar sales in all other states for this period and comprised well over half of the sales for the entire country. In 1869 alone, according to Paul W. Gates, 1,726,-794 acres were sold by the federal government in California. In a twenty-year period after 1862, over 7,000,000 acres of federally owned land was purchased either with warrants or scrip or for cash.

One of the most notorious land speculators of this period was William S. Chapman of whom it was said that "land officers, judges, local legislators, officials in the Department of Interior, and even higher dignitaries were ready and anxious to do him favors, frequently of no mean significance." Between 1868 and 1871, Chapman entered at federal land offices 650,000 acres of land in California and Nevada. At the same time, according to Mr. Gates, he entered additional lands through dummy entrymen, purchased many thousands of acres of so-called "swamp" lands from the state, and otherwise added to his holdings until they totaled over 1,000,000 acres. "Fraud, bribery, false swearing, forgery, and other crimes were charged against him but he passed them off with little trouble." When his holdings are plotted on a land-use map, it appears that he had acquired some of the choicest and most valuable farm lands of the state. Chapman was the largest but not the only land speculator of the period. Mr. Gates, in fact, lists 43 other large operators, who, in the sixties, had acquired 905,000 acres. Eventually most of this land was resold but at inflated prices, and only after the initial wave of bona fide settlers had broken against the wall of land monopoly in California. "Buying in advance of settlement," writes Mr. Gates, "these men were virtually thwarting the Homestead Law in California where, because of the enormous monopolization (through speculative purchases), homesteaders later were able to find little good land."

To appreciate the "land squeeze" felt by American farm settlers in California, it should also be kept in mind that, by 1880, the railroads had received patents to 11,458,212 acres. Approximately 16 per cent of the entire land area owned by the federal government in California—land which might otherwise have been open for free settlement—was given to the railroads. As late as 1919, the Southern Pacific Company was still the chief landowner in the state, with 2,596,775 acres in Southern California alone, including 642,246 acres in one county. To be sure, much of the railroad land was of little value, notably the alternate sections in the desert areas; but the Southern Pacific Company at one time owned some of the finest lands in the San Joaquin Valley and its holdings also included valuable timber and oil properties. In fact, it still owns large and valuable tracts in the Central Valley.

When the Southern Pacific began to extend its line through the valley in the seventies, it encouraged settlers to occupy its lands upon the representation that, later on, "moderate" prices would be fixed and sales agreements would be concluded. The promotional literature issued by the company made vague references to prices of $2.50 and $5 an acre, with $10 being the highest price mentioned. Later these settlers discovered that the lands they had occupied were being placed on the open market for sale to the highest bidder; that prices had been fixed which ranged from $25 to $40 an acre; and that, included in the sale, was every improvement which they had made upon the land—orchards, crops, irrigation systems. It was this situation that touched off the famous "Battle of Mussel Slough" in which five farmers were killed in resisting eviction orders issued by the company and many more were arrested. When the five dead farmers were buried on May 12, 1880, writes Oscar Lewis, "a funeral queue two miles long followed the hearses to the cemetery." Twenty years later this incident was eloquently described in Frank Norris' famous novel, *The Octopus*.

By an act passed in 1850, Congress granted to the public land states all the "swamp and overflow" lands within their boundaries. This act came to the attention of Congress "as a meek innocent-looking stranger." It seemed fair to grant swamp and overflow lands to the states since it was said that these lands were of "no earthly value." As a matter of fact, this statement was true of the swamp and overflow lands in most of the western states; but it was not true of those in California. The abuse of the act in the other western states, moreover, was kept at a minimum since, in these states, a swamp was a swamp and the word "overflow" had a fairly precise meaning. But this was not, of course, the case in California. Many large tracts in California, which had the appearance of being swamp and overflow lands during the rainy season, could actually be cultivated for nine or ten months of the year without drains or levees or other reclamation work. Much of the land surveyed as "swamp and overflow" in California had, as Horace Greeley put it, "not muck enough on the surface to accommodate a single fair-sized frog."

Up to 1907, 2,042,214 acres had been turned over by the federal to the state government as "swamp and overflow" land. These lands, of course, were automatically removed from the operation of the Homestead Act. Once the state got title, it proceeded to sell the lands for $1.15 and $1.25 an acre. Included within this princely federal grant, so improvidently disposed by the state, were some of the richest farm lands in California. The disposal of these lands was accompanied, as an investigating committee later discovered, by the "grossest fraud." Through the connivance of "various parties," surveyors were appointed who "segregated lands as swamp which were not so in fact." For example, Henry Miller, the great land baron of the period, hitched teams of horses to a boat and had himself pulled about over a vast tract so that he could later swear that the land belonged in the "swamp and overflow" category.

Somewhat similar conditions prevailed in other states which received swamp and overflow grants but the peculiar state of affairs in California allowed widespread fraud and chicanery.

In addition to swamp and overflow lands, the state of California was granted 6,719,324 acres from the federal government for schools. This land, of course, was also removed from the operation of the Homestead Act. By 1869 most of the school lands had been sold, for a pittance, and hardly any of the swamp and overflow lands remained in state ownership. "Thus in eighteen years," wrote Bancroft, "the state had disposed of her vast landed possession." To show how unimportant the Homestead and Pre-emption laws were in California, one need merely summarize certain facts: 8,850,143 acres were confirmed to the holders of land grants; 8,426,380 acres granted to the state for various purposes had been largely disposed of by 1880; 7,000,000 acres of federal land had, by the same year, been purchased with cash, warrants or scrip; and 11,458,212 acres had been patented to the railroads. All in all, these dispositions amounted to over 35,000,000 acres or well over one third of the total area of the state. It is apparent, therefore, that only a very small acreage of land in California was ever entered by bona fide homesteaders. Although a somewhat similar pattern could be traced out, in the other western states, the appropriation of the public domain was achieved in California with a "peculiar directness": not by stealth and indirection, over a period of years, but by force and fraud at the beginning of statehood.

A bare factual statement of what happened to the public domain, however, fails to convey the reality of early-day land monopolization in California. By merely restating certain salient aspects of the career of Henry Miller, "The Clemenceau of the Plains," it is possible to convey an understanding of the processes involved. Born in Germany, Henry Miller came to California in 1850. When he arrived in San Francisco, he had precisely six dollars in his pocket. In the flush times of the period, he was soon able to get a start in the butcher business in which he had had some prior experience. In the course of making excursions into the San Joaquin Valley to buy cattle, Miller became interested in, and eventually purchased, a number of Mexican grants at prices as low as $1.15 an acre. In this way he acquired, in a short period, Rancho Santa Rita (48,000 acres), Buri Buri, Salispuedes, Juristac, La Laguna, Aromitas y Agua Caliente, San Antonio, San Lorenzo, Orestimba, Las Animas, Tesquesquito, and other grants. One of the devices which Miller used in acquiring these grants was to purchase the rights of one of several Mexican heirs. Ownership of this interest gave him the right, as a tenant in common, to graze cattle over the entire grant. In most cases he was so ruthless in abusing this right that it was not long before the remaining heirs were willing to sell their interests to him for a nominal price. In addition to grants, he acquired over 180,000 acres of federal land, most of which was purchased with depreciated scrip and warrant certificates which he had picked up for a song. Eventually Miller & Lux owned an empire in California as large as Belgium, embracing 1,000,000 acres, and it was Henry Miller's boast that he could ride from Oregon to Mexico and sleep every night in one of his ranches.

One of the means by which Miller was able to acquire this vast domain was through the unscrupulous use of riparian water rights to force other landowners to sell their holdings to him. Making strategic acquisitions of land, with an eye to riparian rights, he used these rights to monopolize the water supply. Many of his holdings, also, were in the trough of the valley, and thus acquired the benefit of water by drainage which other owners had brought to their lands at great expense. This strategic use of water monopoly, in fact, was one of the principal means by which the early land monopolies were built up and kept intact long after the time when, normally, population pressure would have induced a subdivision of large holdings. Through a system of loans to county assessors and other officials, Miller was also able, over a long period of years, to keep the taxes on his holdings extremely low. Eventually the Miller & Lux holdings changed hands and a great agricultural empire exists on these holdings today; but the subdivision and resale was delayed long enough to make it possible for Miller and his associates to make enormous speculative profits.

A further factor that entered into land monopolization and kept large holdings off the market for many years was the availability, in California, of a large pool of cheap labor, made up, originally, of Indians, later of Chinese, Mexicans, and other groups. With other costs being approximately equal, the large-sized farm unit in California was able to keep costs lower than the family-sized farm by using a large supply of cheap, migratory labor which could be utilized only when needed. The nature of the terrain, also, made large-scale operations more feasible than in other areas. As profits were turned back into development and irrigation projects, land values were capitalized at ever-larger figures which, in turn, made it difficult for the farm family to acquire land. Eventually most of the large holdings were reduced in size or broken up but not until land values had been capitalized at an exceptionally high level and, in some cases, the original holdings are still more or less intact.

The points to note about land monopolization in California are, first, the "peculiar directness" with which land was monopolized; and, second, the abnormal delay in the breaking up of large holdings. The delay in the subdivision of these holdings largely defeated the brilliant promise of the first two decades after the discovery of gold. By 1870 a general dry-rot had set in. "The whole country is poverty-stricken," wrote one observer; "the farmers are shiftless and crazy on wheat. I have seen farms cropped for eighteen years with wheat, and not a vine, shrub or flower on the place. The roads are too wide, and are unworked, and a nest for noxious weeds. The effect of going through California is to make you wish to leave it, if you are poor and want to farm." Considering that California was theoretically still a "frontier" state in 1871, it is interesting to find Henry George writing in that year that "California is not a country of farms but a country of plantations and estates. Agriculture is speculation. . . . There is no state in the Union in which settlers in good faith have been so persecuted, so robbed, as in California. Men have grown rich, and men still make a regular business of blackmailing settlers upon public land, of appropriating their homes, and this by power of the law and in the name of justice." Stephen Powers, who tramped through rural California in 1872, wrote that he had not seen "ten honest, hard-fisted farmers in my whole journey. There are plenty of city-haunting old bachelors and libertines, who own great ranches and lease them; and there are enough crammers of wheat, crammers of beans, crammers of mulberries, crammers of anything that will make their fortune in a year or two, and permit them to go and live and die in 'Frisco. Then, for laborers, there are runaway sailors, reformed street thieves, bankrupt German scene-painters, who carry sixty pounds of blankets, old soldiers who drink their employers' whiskey and fall into the ditch which they dug for a fence row." It was not unlikely, Powers thought, that "within two centuries" California would have a division of population something like that of ancient Greece: "great lords of the soil," on the one hand, and "a kind of peasantry" on the other. This prediction was fulfilled but within two decades, not two centuries. Here, again, one notes the telescoping of economic and social processes which is so characteristic of California. Phases in the process of land monopolization which have taken centuries in Europe have been repeated in California in the course of three generations.

THE MEANING OF LAND MONOPOLY

One of the most striking respects in which California differs from the other western states consists in the manner by which it skipped or omitted the frontier phase of land settlement. California began with land monopoly and, in this respect, it is an exception to the rule of frontier settlement. "Ever since American rule superseded Mexican," writes Dr. Paul S. Taylor, "California has found its agriculture lying outside the American farm tradition. Skill, organization, development, magnitude, all these have been worked into her State agricultural history. But, *unlike the nation,* California has not placed at the head of her agricultural goals the achievement of a satisfactory relation between land and the families who labor upon it." Only to the degree that one can appreciate what the American farm tradition has meant in other states is it possible to understand the peculiarity, the exceptionalism, of California in this respect. It is the one state in the Union in which the American farm tradition has never existed, except in a most limited and never fully realized manner. It would, perhaps, be more accurate to say that California has always had a unique farm tradition which is made up of several different elements: the tradition of the large Spanish hacienda; elements of the Southern plantation tradition and the self-sufficient small farm of the North and Middle West (since elements of both traditions are present in California); and an element which stems from early western mining. Apart from the specific contributions which mining made to agriculture in California, farming has always resembled mining in this state. The soil is really mined, not farmed.

From this difference in the character of the farm tradition in California, certain exceptionally important social consequences have issued. The more important of these consequences can be summarized as follows: a marked degree of social instability; the existence of a long-standing, and cancerous farm labor problem; the continued existence of an unresolved land problem; and the development of a most exceptional social structure. The social instability is shown by the fact that the people of California have repeatedly sought to change the basic relationship of the people to the land. Over a long period of years, they have sought to bring this relationship into a more general conformity with the American farm tradition. In fact, they are still trying to effect this modification. A large part of the social history of the state deals with various phases of this century-old struggle: the struggle to exclude slave labor; the fight to exclude Oriental immigration; the various movements to curb land monopoly; the long-continued and embittered fight to break the water monopoly; and the current fight to retain the 160-acre limitation on the Central Valley Project.

Down through the years, the rural areas of California have echoed with various forms of social conflict which are strikingly at variance with the traditional picture of rural life elsewhere in America: squatter riots; night riders; rancorous litigation over water rights; armed resistance to court orders; and similar noises, disturbances, and conflicts. The existence of an unresolved land tenure problem is one of the basic explanations for the unusually high concentration of population in urban areas in California. Many observers in the sixties and seventies were struck by the fact that half the population of California was concentrated in San Francisco in the winter months, and by the exceptional size of the landless, rootless, floating character of a large section of the state's population. This existence of a large landless population and a highly urbanized population account for, at least in part, the marked social instability which has always characterized California.

California's divergence from the national farm tradition is also reflected in the remarkable way in which urbanism, as a way of life, has invaded rural areas. This invasion, in turn, has brought into being a type of social structure in rural areas that does not have its precise counterpart in any other state. One can find in Dr. Walter Goldschmidt's excellent volume, *As You Sow,* a most systematic and scientific description of this anomalous social structure. Dr. Goldschmidt finds, for example, that "rural" life in California is more urban than rural. The population is as heterogeneous in origin as the typical urban community and shows much the same diversity of social action. As he so well points out, "the lifeways of a potato grower" in California are unlike those of a grape producer, and much the same specialization is to be found among farm laborers. Pecuniary standards dominate the rural scene in California. Production for family use is negligible. "Sharing of implements and trading labor are so rare as to appear unique." The farm population is orientated toward the market and both farm and small town reflect the "hierarchy of elites" to be found in urban centers. "Rural slums" are to be found throughout the state. The California farm-family is small. The traditional "hired man" has disappeared and the society is sharply segmented along economic class lines. Even the "towns" are not like farm centers elsewhere. The assumption that a dichotomy exists between town and country, urban and rural areas, has determined, as Dr. Goldschmidt points out, various forms of public action in this country, such, for example, as the form of county government which caters to "rural" needs. When forms of public action, based on this assumption, are applied in California, they make for maladjustment, since they do not fit the reality. And this maladjustment, in turn, contributes to the general social and political and economic instability.

For an understanding of many phases of the California riddle Lord Bryce's observation is still pertinent: "Latifundia perdunt California."

[7]

THE MOSAIC OF CALIFORNIA'S AGRICULTURE

◇◇◇

THOUSANDS OF American motorists have voiced protests against California's notorious plant quarantine inspectors who vigilantly guard the borders of the state to prevent the importation of certain insects and crop-destroyers. No other state maintains a comparable service. But there is a good reason for California's vigilance and its border inspectors might well be taken as symbols of the exceptionalism, the peculiarity, of the state's agriculture. For these inspectors are guarding an agricultural production as fabulously rich as it is utterly unique. There is no parallel for the state's agriculture, either in this country or in the world, and, since agriculture has long been a basic industry, it is quite apparent that the uniqueness of California's agricultural pattern has had a profound influence on the state. California's agriculture is a wonderfully intricate and novel mosaic. It is the purpose of this chapter to describe the parts of this mosaic and how they fit together to make up the exceptional design.

HER INFINITE VARIETY

The most obvious fact about California's agriculture is its extraordinary diversity. Today the state produces 214 different agricultural products, including some 35 field crops, 68 fruits, 86 vegetables, and 40 different commercial live-stock, poultry, and honeybee products. The extent of this diversity can best be measured by comparing the range of California's production with that of typical corn, cotton, or dairy belt farming areas in other states which may produce from 12 to 15 different crops. The diversity of California agriculture can also be measured by noting the variety or types of farming which exist in the state. Approximately 118 different and distinct types of farming can be found in California by comparison with 8 types to be found in Illinois, 12 in Kansas, 20 in Texas, and 25 in Pennsylvania which, in this respect, ranks second to California. Here the disparity, 118 to 25, is some index of the amazing diversity of California's production.

The diversity of California's agriculture is, of course, a direct reflection of the amazing range of environmental factors to be found in the state. California has the highest peaks, the lowest valleys, the driest desert, and some of the rainiest sections of the United States. The variety of soils is so great that California is one of the world's great laboratories for the study of soil-forming and soil-reacting processes. "For many decades," writes Dr. Hans Jenny, "the soils of California seemingly did not fit into the prevailing scientific systems of soil classification." As late as 1935 scientists were unable to delineate in California representative types of the great climatic soil groups of the world; later these types were found but in unique combinations and freakish variations. The climatic environment has two unusual characteristics: a wide range in total annual precipitation (from almost none to more than 100 inches of rainfall); and a most unequal seasonal distribution between a short rainy season and a long growing season. "Many of the warm-season crops," writes Dr. Jenny, "find suitable climatic conditions available at some time in the year almost everywhere in the state; and conversely, there is no season when they cannot be produced in some district." This amazing range of climatic conditions not only greatly extends the growing season for many crops but it also makes for an extraordinary variety of production. Fruit ripens first in the *northern* part of the state, reversing the usual pattern. The topography is, also, exceptional. There is very little level land along the coast for the coast ranges rise near the shore; but the Central Valley, 600 miles long, is nearly dead-level, since the Sierras drop sharply on the western side and the streams have not had time to form valleys. In some areas, the drainage is away from, not toward, the rivers. One can sum up the diversity of the environment by saying that in no other area of the world, of comparable size, is so wide a variety of ecological factors to be found.

One way to bring out the uniqueness of California's environment is to compare it with the world-region which it most closely resembles—the Mediterranean world of Italy, Algeria, Tunis, Sicily and Greece. California and Italy are about the same size; they have much the same type of Mediterranean climate; and they do not differ greatly in the amount of rainfall. Yet, despite this general similarity, the differences are perhaps more significant than the resemblances. If California really resembled Italy, then the value of its crops per acre would decline from north-to-south; but, unlike Italy, the southern coastal plains of California have some of the heaviest crop-yields to be found in the state. "The height and location of the California mountains," writes Dr. Ellsworth Huntington, "and the relatively low temperature along the coast alter the distribution of productivity to the great advantage of the southern two-thirds of California. The mountains, especially the lofty Sierras with their snows which last far into the summer, make it possible for California to have a higher percentage of irrigated land than any other state. . . . The low temperature along the coast prevents the dry summer from injuring vegetation as it does in Italy. It also enables a given amount of water to irrigate a larger area than in the hotter interior." California's areas of greatest productivity would be, in corresponding areas in Italy, areas of lowest productivity.

Reflecting this amazing topographical, climatic, and soil diversity, California's agriculture is perhaps the most diversified and varied to be found in the world. Were it not for the remoteness from markets, California could grow almost any crop on a competitive basis; but, even so, its list of agricultural products represents a fair sample of the world's agriculture. California is said to be weak in corn, cotton, and wheat, but, at one time, it was one of the world's great wheat-producing areas; it could produce corn if non-environmental factors were more favorable; and, in recent years, cotton has become its main agricultural income producer. In fact, California now ranks fifth among the cotton-producing states.

[*105*]

CALIFORNIA: *The Great Exception*

AMERICA'S NATURAL HOTHOUSE

The second quality to note about California's agriculture is its extraordinarily dynamic quality. Farm production has increased faster in California than in any other state. By 1924 it had joined the elite of the farm-producing states; by 1929 it had passed Texas; in 1930 it held first place; and, in 1949, it ranked second to Iowa, with annual cash farm receipts totaling $2,207,639,000. Not only has its agriculture shown an amazing quantitative increase, but its various metamorphoses have been equally amazing. In the period from 1848 to 1948, the agriculture of the state has passed through "all the stages exemplified by several centuries of the world's agricultural history." Each major phase of this world-history has been repeated but with the time-span being greatly foreshortened. In this sense, California agriculture has a dynamic quality which, in the words of one expert, has seldom, if ever, been duplicated. The basic explanation for this extraordinarily dynamic quality is to be found, of course, in the fact that California is a great natural hothouse, air-conditioned and air-controlled, in which growth can be greatly accelerated in time and in quantity. But there are other reasons, social and historical, which must also be taken into account.

Although little agricultural development had taken place in California prior to 1848, some extremely important precedents had been established which greatly facilitated the enormous expansion which took place under American rule. The area farmed by the Franciscans was not large, perhaps not more than 10,000 acres, but the Mission "gardens" were the proving-grounds for many crops. "The fruits and nuts known to have been grown by the padres," writes Dr. Frank Adams, "included almost all those now produced in California, and some that have not succeeded commercially. There were pears, peaches, apples, almonds, plums, quinces, pomegranates, oranges, lemons, citrons, limes, dates, cherries, plantains, walnuts, grapes, olives, figs, strawberries, and raspberries. Even this list is not complete." Some of the fruits now grown in California are direct, lineal descendants of fruits planted in the Mission gardens, notably, the Mission grape, Mission fig, and Mission olive.

It was a happy circumstance that California was first settled by immigrants from Spain, for the seeds that they brought from Spain and Mexico were well adapted to the environment; and, generally speaking, the type of agriculture which had its miniature beginnings in the Mission gardens was quite unlike that with which American farmers were then familiar. Although this Mission production was little more than a memory in 1848, it later became extremely important. The Franciscans were, also, familiar with irrigation, a new art to the first American immigrants, and they made some important demonstrations of the feasibility of irrigated farming. In short this experimental "development-in-miniature" was an important cultural factor in the swift expansion of agriculture which took place after 1848.

Then, too, the discovery of gold created overnight the purchasing power and the markets for a rapid expansion of agriculture. Fabulous prices were paid for agricultural products, particularly for fruits. The owner of a single peach tree in Coloma, which produced 450 peaches, sold these peaches in the mines for $3 each. An orchard of fifty trees produced $2,800 for the owner in a single year. The combination of people plus gold acted as a most powerful dynamic in forcing a rapid expansion of agricultural production and, in the years since the gold rush, an ever-increasing tide of migration has continued this dynamic. Moreover, mining was related to the mushroom-growth of California agriculture in other respects. For example, flumes constructed by miners were later used as rudimentary irrigation works and many miners used their earnings to purchase farms and to develop new lands.

A third factor is to be found in the fact that the discovery of the amazing versatility of the California environment happened to coincide with national developments which provided still further dynamics for the expansion of agriculture. The sharpest percentage increase in the urban population of the nation took place in the years from 1850 to 1870. The decade ending in 1860 witnessed a 92.1 per cent increase in the urban population and the next decade, which saw the completion of the transcontinental railroad, witnessed a 59.3 per cent increase. The rapid increase in the urban population and the completion of the rail line were largely responsible for the swift transition from wheat farming to fruit-and-vegetable farming which took place in California in the 1870's. Just as an increasing proportion of the American people began to get their food from grocery stores, and not from gardens and farms, one of the world's great "gardens" was discovered in California.

A final factor, underlying the remarkable dynamics of California agriculture, consists in the availability of a large supply of mobile labor. From the earliest years, California farmers have drawn their labor from a general pool, using a large supply when needed, and releasing this supply as soon as a particular job was completed. The availability of this labor supply made possible not only the rapid expansion of agriculture, but its remarkable specialization. There are, of course, still other factors which serve to account for the dynamic quality of California agriculture but most of these additional factors have to do with cultural practices and environmental conditions which will be considered later in this chapter.

THE MEANING OF SPECIALIZATION

Still another extraordinary quality about California agriculture is its high degree of specialization. Only 6 per cent of the state's 132,658 farms are so-called "general" farms; the others are all specialized in production. With only 9 per cent of its population living on farms, California was able to produce an agricultural income of more than two billion dollars in 1948. Just as diverse environmental factors have made for diverse production, so diversity of production is the parent of specialization. That California could produce so many crops that could not be produced elsewhere naturally made for specialization; but specialization has several different dimensions as it relates to California agriculture.

First of all, California agriculture shows a remarkable specialization-by-area. This type of specialization did not come about overnight; it was the end-product of long, costly, wasteful, and painful experimentation in a novel environment. Since the environment is highly versatile, many of the state's basic crops were at first planted on a state-wide basis. The fact that these crops could grow, rather than how well they grew, was the important consideration. But, after years of disastrous experimentation, it was discovered that certain crops could only be produced on a commercially successful basis in particular areas. In this sense, the development of California agriculture has been an enormously costly and destructive process; almost as many orchards have been ripped out as have been planted. A degree of stability has been achieved only as the areas of specialized production have been isolated and identified.

Almost every crop now grown in California has had to find its special area. Almond production, once widely dispersed, is now largely concentrated in portions of two counties. In the 1860's bearing apple trees were reported in nearly every county; today commercial production is concentrated in two districts. Grape production, once concentrated in the south, shifted to the central part of the state in the 1890's. Apricots, peaches, and cherries, once grown nearly everywhere, are now highly concentrated in a limited number of areas. Within this larger specialization-by-area, many further degrees of specialization will be noted. For example, the Watsonville area specializes in the production of winter apples; the Sevastopol area in the production of summer apples. Affected by changing costs, new cultural practices, shifting market demands and other factors, this process of adaptation has been continuous. Entire areas of the state are devoted to the production of a single crop, often to the production of a special variety of this crop.

Specialization-by-area, however, fails to suggest the extent to which California agriculture is specialized. For the same factors which have brought about a specialization by area have also worked toward a specialization in the varieties of particular crops, and in the timing of the maturity dates for these same crops. For example, the Calimyrna fig (California Smyrna) is raised for drying; the Kadota, for canning. Apricots are a three-purpose crop: for the fresh market, for canning, for drying. One variety of peach will be raised for the canning industry; other varieties for the fresh market. This specialization is often reflected, although not necessarily, in area divisions. The production of canning peaches, for example, is concentrated in a few districts. Specialization, in other words, is by area, by variety, by maturity dates, and by function (as, for example, wine grapes, raisin grapes, and table grapes). Table grapes are harvested in the Coachella Valley in June; in the Central Valley from July to September; and, still further north, from September to October. Celery, localized in area, is planted at intervals so as to insure a continuous year-round harvest.

Specialization by area is reflected in the bewildering variety of "festivals" celebrated throughout the year in California. There is the "Potato" festival in Shafter in June; the "Tomato" festival in Niland in January; the "Carrot" festival in Holtville in February; the "Orange" festival in San Bernardino in March; and the "Grape" festival in Fresno in September. Often a community will honor a *particular* variety of a crop, as the "Emperor Grape" festival in Exeter.

Specialization for particular markets is also a striking characteristic of certain phases of California agriculture. An Imperial Valley "shipper-grower," for example, will discover that a certain product commands a premium price on a particular eastern market during a brief interval of time. By the intensive use of soil foods and other cultural practices, he will plant a crop carefully timed, to a matter of days, to reach this *one* market at this *particular* time. A week's delay in maturity or in harvesting will often result in great financial loss. A large part of California's production is of the "off-season" or "early season" variety; in fact the premium prices which this type of product commands offsets the disadvantages of remoteness from the market. Frequently the first section of the crop to mature goes to the "fresh" market locally; the second is used in the "shipping deal"; the third for the canneries. Often production is carefully synchronized between areas. For example, lettuce is shipped from Salinas between March and December, from Imperial Valley between December and March. In this instance, the entire industry, including shipper-growers and lettuce-packers, migrates from one area to the other. To bring off a synchronization of this delicacy requires careful planning, area by area, crop by crop, schedule by schedule; a high degree of organization; and centralized authority and control.

The Mosaic of CALIFORNIA's Agriculture

The same diversity and specialization may be found, also, in the livestock industry in California. Many different breeds of sheep have been found useful in California because of the variations in climate and topography. In fact, there are so many different types of wool produced in the state, varying in grade, length, and yield, that California has been called "an ovine menagerie." Here, again, production is highly specialized. Certain areas concentrate on the production of wool; others on the production of mutton. The earliest milk-fed "springers" to reach the eastern markets are from the Sacramento and San Joaquin Valleys where, because of special climatic conditions, lambs are matured more rapidly than elsewhere.

Dairy farming, which is widely distributed in area, is nevertheless more highly specialized than in any other state. Poultry farming has, likewise, developed on a highly specialized and localized basis. As early as 1904, nine-tenths of the people living in the vicinity of Petaluma were engaged in the poultry industry. Here, also, the key to specialization is to be found in the fact that it was discovered that January, February, and March, which are months of low egg production in the Midwest, are months of high production in California. Poultry raising in the Petaluma area naturally attracted the hatchery industry; there are hatcheries in the area today which can incubate 1,800,000 eggs in a single process. The production of beef cattle was once conducted in California on the extensive scale familiar throughout the Southwest; but, today, it is highly specialized and localized.

The same pattern appears in the floral industry. A truly enormous income is produced in California by highly specialized floral industries. For example, there is a Cymbidium orchid factory at Ojai,—El Rancho Rinconada,—which cost $8,000,000 to build and develop. One plant alone produces around 400 flowers a year which sell for $2 apiece. This particular factory, which was started by the owner as a hobby, now produces nearly one million orchids a year. Like so many similar industries, including a large poinsettia and lilac industry, the orchid industry was located in California by reason of the environment. There are literally dozens of exotic and highly specialized industries of this kind to be found in California and, in the aggregate, they produce a high income and are largely monopolistic in the sense that their products cannot be commercially produced in other areas.

It does not, in fact, make much difference which part of California's mosaic one examines for the same specialization will be found, from poultry to prunes, from tomatoes to orchids. This specialization reflects the diversity and versatility of the environment and the uniqueness of California in the total American scene. If there were other "Californias" in the East or South or Midwest, the specialization which exists in California today might be much less. By reason of its remoteness. from non-local markets, also, California was forced to develop specialty products, and often superior products, to offset the costs of transportation. Specialization, in turn, has made for higher yields in almost every crop, from eggs to cotton. It has also given California a near-monopoly in the production of many crops: olives, lemons, almonds, pomegranates, nectarines; most of the nation's apricots, avocados, artichokes, fresh asparagus, broccoli, lettuce, melons, figs, Persian and English walnuts, persimmons, dates, grapes, plums, and dried prunes. California packs almost all the white and about half the green asparagus canned in the United States and it ranks first in the output of peaches, pears, plums, and oranges. More than 90 per cent of the national production of dried fruits originates in California and the state packs about 25 per cent of the nation's canned fruit. Year in and year out, California produces about 45 per cent of the nation's crop of fruits and nuts.

Specialization, in its various aspects, has been a factor of prime importance in the development of compact organization, both for production and marketing, and in the application of technology. Where most of the producers of a single crop, or of a special variety of one crop, are located in a particular area, it is much easier to organize a cooperative marketing association than if they were scattered over several states. Similarly, given the same degree of specialization, it is much easier to develop and to apply new cultural practices and technological innovations. Specialization simplifies the collective approach to many problems: from pest control to marketing. Essentially specialization in production makes for specialization in attitudes and interests; California producers are not "farmers," they are avocado-producers, citrus-producers, raisin-growers and so forth. Specialization also accounts for the fact that California agriculture is overwhelmingly commercialized, the distinction between "cash" and "non-cash" crops having little meaning. Specialized production has its special perils, pitfalls, and problems but, on balance, California has profited enormously from the highly diversified and amazingly specialized character of its agricultural production.

THE WORLD IN MINIATURE

Specialization has still another, and generally ignored, meaning in relation to California's agriculture. Many different racial and ethnic groups have been attracted to rural areas in California and each group has brought with it some special skill or technique, seed or plant. By and large, these groups have been drawn to California by the diversity of the state's resources for a special niche has been reserved for each of them. In fact it is almost axiomatic that what will grow in the Mediterranean area, the Near East, Europe, or the Far East, will also grow in California. California agriculture has been the recipient of extremely important cultural contributions from the Chinese, Japanese, French, Basques, Mexicans, Armenians, Slavs, Portuguese, and many other groups. Much of the traditional lore and many of the skills which these groups have brought to California, being oral in nature, could only have come with the people themselves. Rural California is still dotted with racial and ethnic colonies and the diversity of the rural population matches or parallels the diversity in production.

Apart from the long list of agricultural products which the Spanish demonstrated could be grown in California, the state got its Double Dwarf milo maize from Japan; alfalfa came from Chile; a variety of the lima bean, widely planted in the state, is derived from a seed developed by the Hopi Indians; its flax came from India; the avocado and tomato from Mexico; its dates from Algeria, Egypt, and Persia; its figs from Smyrna; the Goleta walnut from Chile; an important variety of plum from Japan; certain varieties of pears from China, and prunes from France. Over twenty different nationalities have made contributions, at different times, to the development of the grape and wine industries of the state. In fact, viticulture in California owes a lasting debt to French and Hungarian vineyardists who settled in the state. The produce industry in California owes an enormous debt to Japanese farmers. The Chinese played a key role, not only as laborers in the orchards, drying-sheds, and canneries, and in the building of the dikes and levees in the Delta section, but in the development of many crops, notably celery. Slavonians were pioneers in the apple industry in the Pajaro Valley which they still largely control. The Portuguese are a powerful element in the dairy industry in the San Joaquin Valley. Henry Markarian was one of the first fig growers in the Fresno area, and the first fig packing plant was established by the Seropian Brothers. Chinese immigrants set out the first pear orchards in the Sierra foothills, and a Chinese farmer made the first shipment of potatoes from Kern County (an area that produced 960,000,000 pounds of potatoes in 1944!). Indeed, it would be difficult to exaggerate the importance of the contribution which these varied immigrants have made to the mosaic of California agriculture.

Much of this cultural history has been lost but enough material exists to demonstrate how important the contributions of these immigrant groups have been. Commenting on the extraordinary increase in rice production in California—from 70,000 bushels in 1912 to 9,000,000 bushels in 1919—one historian has said that an increase of this magnitude in one decade is "too great to be either fully understood or appreciated." But Oriental immigrants in California could provide the answer. As early as 1865, 20,000 pounds of rice were imported by Chinese merchants to supply the needs of Chinese immigrants and, out of this need, with the skill and knowledge of both Chinese and Japanese farmers, came California's rice industry. Not only did Oriental immigrants pioneer in the production of this crop, working as laborers in building the dikes and irrigation systems used in flooding the rice fields, but for many years three-fourths of the California rice production has been exported to Hawaii for sale to other Orientals.

In the early 1890's, one Juan Murrieta of Los Angeles, imported a variety of avocado trees from Atlixco, Mexico, and from this group of seedling trees came the varieties that were first planted for commercial production. It was in Atlixco that scientists in 1911 discovered the Fuerte variety; today 85 per cent of the trees in Southern California's 16,000 acres of avocado orchards are of this variety (21,300 tons of avocados were produced in 1944). For many years, nurserymen used to visit the fruit stands along Main Street in Los Angeles where Mexican seedling avocados were sold, chiefly to the Mexican population, to purchase seeds of avocados that decayed before they could be marketed. From these seeds the nurserymen grew wild Mexican rootstocks upon which they budded the best-known varieties of avocados. On March 8, 1949, a delegation of Southern California avocado growers made a junket to Atlixco to plant an avocado tree in the plaza of the village as a gesture of thanks and appreciation. Among the first nurserymen and importers of ornamental plants, shrubs, and flowers, who came to California during the gold rush were Bernard S. Fox of Ireland, Stephen Noland from England, Louis Prevost from France, and Frank and Peter Kunz from Germany. "Their willingness to try almost anything," writes one historian, "and their knowledge of practices abroad, greatly benefited early ornamental horticulture." Christopher Colombo Brevidero, an Italian immigrant, is the founder of the important lilac industry in Southern California.

Today an important tomato industry exists in the vicinity of Merced, California, employing a thousand people, which in 1948 produced $3,138,278 in income. The founder of this industry is one Camilla Pregno, an Italian immigrant, who in 1900 taught the Merced farmers how to grow tomatoes on stakes after the manner he had learned in Italy. By this method, 10,000 plants may be produced for each acre, whereas, under the former style of open planting, where the plants were permitted to lie on the ground, only about 1,500 plants can be produced on an acre. Today 75 per cent of the acreage is planted to staked tomatoes.

In the vicinity of the towns of Paramount, Artesia, Clearwater, Bellflower, and Norwalk, in southern California, is a dairy industry which produces 500,000 gallons of milk a month and yields

$61,000,000 annually. This industry has been built up largely by Dutch immigrants who started settling in the area at the time of the first World War. Over the years, new immigrants have arrived and are still arriving so that there are now approximately 20,000 Hollanders in the area. The industry is one more example of the forced-growth technique which is so widespread in all forms of California livestock and agricultural production. Forced feeding, the so-called "shotgun" technique, is used throughout the industry. Dairy cows are fed copra, imported from the Philippines, linseed meal, cottonseed meal (from the San Joaquin Valley), and hay. The average dairyman has 100 cows and on 10 acres or less of land, and expects each cow to produce 1,200 gallons of milk a year. As fast as cows are "burned out" by this method, they are sold for beef and new cows are purchased. This is not, of course, farming; it is a form of mining. The cows are in stalls and corrals, for there are no placid meadows and land is too costly for alfalfa growing. Paved streets, instead of country roads, bisect the even rows of houses and barns and the rumble of feed trucks is never stilled. Dutch immigrants, with a substantial number of Swiss and Portuguese dairy-farmers, have built up this amazing industry.

CHALLENGE AND RESPONSE

The discovery of the agricultural richness of California and the long struggle to unlock this richness make one of the most fascinating stories in the history of the transfusion of cultures. When the early settlers turned from mining to agriculture, they were confronted with a hundred novel and challenging problems. It did not take them long to discover that the "barrenness" of California was a myth. They found that nearly anything would grow; but there were no guides by which they could chart a course or find their way through the freakish upside-down environment they had to master. There were no soil maps; no manuals on pest control or cultural practices; no treatises on irrigation. Yet everything about farming in California was different from the farming that they had known in other areas.

There was, first of all, the basic problem of developing a working knowledge of soils in California. Such knowledge of soils as the settlers brought to California was based upon experience in the humid East and parts of Western Europe and, as Dr. Hans Jenny has pointed out, "often proved of questionable value." A systematic inventory of soil resources was not undertaken until 1913 and, at the outset, there were, of course, no soil maps. Standards of good soil, moreover, hardly apply in California. Many productive citrus, fig and olive orchards are found on iron-hardpan soil which had to be blasted with dynamite before the first trees could be planted. Contrasts in climate and topography have an important relevance to soil uses and nowhere are these contrasts greater than in California. Tillage and plowing techniques, developed in other areas, had to be discarded or modified. The problem of soil erosion, as Dr. Jenny points out, differs in California from other parts of the country, and this, too, presented a special problem. Although certain soils were amazingly rich, it was years before the farmers of California realized that the soils of the state are generally deficient in nitrogen. Many of the basic problems of soil analysis in California, in fact, have not been solved to this day. The novelty of many of the problems encountered and of the unexpected findings which have been made constitute a basic challenge to soil scientists. California is a unique laboratory for the study of soil formations and soil-reacting and soil-building processes so that it is not by chance that E. W. Hilgard, who helped found the College of Agriculture at the University of California, should be recognized throughout the world as one of the founders of soil science.

Closely related to the problem of soils was the host of problems that arose in connection with irrigation practices. William Hammond Hall, the first State Engineer in California, once pointed out that never before had such a unique situation existed as that which prevailed in California. Thousands of settlers had poured into a state, in a brief period of time, "hardly any of whom had the slightest idea of water rights systems or irrigation customs and legal and administrative practices." Furthermore the circumstances were such that "probably no other country ever experienced the influence of such strong inducement to the diversion of water" since the stakes were so high and the possibilities so great. For decades California farmers floundered around in a morass of confusion and uncertainty in an effort to apply laws and customs which had no relevance to an arid environment. The system that finally emerged from this struggle was based upon a fusion of the common law of England, the traditions of southern Europe and Mexico, and the mining rules and customs. To this day, the water system of California has no precise parallel in any other state or country. More was involved on the score of irrigation than the adaptation of alien laws and customs, for cultural practices had to be evolved by a trial and error process, to meet the challenge of utterly novel conditions. Answers had to be found to such problems as when and how to irrigate; water formulas had to be worked out for a wide variety of crops, soils, and climatic conditions. Too much water, it was found, could be as harmful as no water at all. Again, it is not by chance that California should have been the laboratory of the West in which irrigation laws, customs, and cultural practices were worked out and later applied, not only in many other Western states, but in many parts of the world. Such institutions as the mutual water company and the irrigation district, for example, had their origin in California.

The preponderance of annual species in California forage created a host of problems for cattlemen and sheepmen who were familiar with a type of forage made up largely of perennials. It took these men years to discover that the typical California forage, which provides a complete diet in the spring, has by fall become deficient in many properties. Livestock was supposed to subsist on natural vegetation whether it had any food value or not; and so the livestock industries were, over a period of years, pushed into the foothill and marginal areas and the state suffered from a deficiency of livestock production. It was not until these livestock men, through the use of empirical methods, discovered that the information which they had brought from other parts of the country was largely worthless in California that a reversal began to occur in the livestock industries. Once their minds were freed from the weight of a vast accumulation of misinformation, these same livestock men began to use experimental methods consciously and with amazing results. California cattlegrowers today use a bewildering variety of feed to supplement natural forage, including cottonseed meal, flax, sugar beets, rice, copra, fish oils, and many other products. The development of new uses for these products has, in turn, stimulated many other industries. In the evolution—one might well say the "revolution"—which has taken place in the livestock industries in California, it is possible to see the outline of a process which has operated throughout the entire range of the state's agriculture, and with amazing results. It is the discarding of traditional cultural practices by empirical observation and study; and, later, conscious experimentation based upon organized research, leading to the adoption of new practices.

In the development of the orchard crops in California growers were compelled, again by empirical methods, to discard many practices which they had learned in other areas. Literally hundreds of different varieties were imported and, since they would all grow in most of the orchard districts in the state, it took years of experimentation to select the types best adapted to the environment. Fifty varieties of cherries were tried before two were finally selected as best adapted; dozens of varieties of pears were introduced, but today one variety makes up 85 per cent of the total acreage; fifty varieties of plums were used, but only a few have survived; and so it goes with almost every orchard crop. Entirely new methods had to be worked out for a wide range of cultural practices, including spraying, fumigating, irrigating, pruning, girdling, packing, curing, thinning, precooling, planting, caprification, air-drainage, canning, pollination, hybridization, and processing. Tricked by the versatility of the environment, growers had to learn all the freakish perils and perverse advantages of this paradoxical country. They had to discover all about the vagaries of fog and frost, underground rivers, thermal air belts, and scattered soils. In the process of doing so, however, they pioneered in the development of new methods and practices which constantly stepped up California production and which, incidentally, have been widely used in other areas. So far as all forms of cultural practices are concerned, California is one vast experimental laboratory, perhaps the largest and certainly one of the most useful agricultural laboratories in the world.

In no other area have so many ingenious and man-made devices been invented to cope with an astonishing range of ecological factors. In the sixties and seventies California farmers experimented with steam-powered tractors, wheel plows, gang plows, rotary spades, screw pulverizers, and many similar devices, most of which were developed to cope with special conditions. Stockton gangplows, Fresno scrapers, Randall harrows, and many types of irrigation equipment had their origin in California. A treatise could be written about the many inventions, innovations, and adaptations of farm implements worked out by Stockton Berry, who developed the first steam combine harvester. In fact it would be impossible even to list the more important farm implements and special devices that have been developed in California, for the list is still expanding so rapidly that no one can keep abreast of the developments. From the current news one can select such items as the following: hydraulically-operated platforms for pruning orchard trees; wind machines for frost protection; electronic sorting devices which sort peaches and other fruits by color; special portable refrigerator units; new portable viners which harvest and shell lima beans; an olive-pitting machine that pits 750 olives a minute (it formerly took an experienced pitter a full day to pit this many olives); hydraulic lifts used in picking dates; mobile telephone services for farmers which transmit calls to and from moving vehicles; special helicopters used in crop dusting, seeding, and spraying; and so on. "Farming from the skies" is largely a California innovation. There are 75 companies in the state which offer airplane services for planting and spraying and fertilizing crops. Almost all the rice crop is now planted from the air and the airplane is in process of becoming an efficient all-purpose farm instrument. The shipment of perishable products by air is still in its infancy but, in 1948, some 13 tons of California figs were shipped to eastern markets by air.

California agriculture is, in almost every field, the most thoroughly mechanized segment of American agriculture. Most seeding, planting, and cultivating operations have long since been thoroughly mechanized and mechanization is proceeding rapidly in the harvesting of even specialty crops. Certain crops—rice is one— are now completely mechanized and remarkable progress has been made in the mechanization of the cotton and sugar beet crops. Pneumatic tree-shakers are in use to harvest almonds, and onion and tomato harvesters are being perfected. California has the most intensive rural electrification program of any area of the world of comparable size. California farmers use more than one-half of all rural electricity used in the United States, with some farm areas of the state being 96 per cent electrified. There are some 37,000 electrically-operated pumping plants in the San Joaquin Valley where the high average use of electricity, despite monopoly controls, has kept power rates at a comparatively low level.

An environment as freakish as that of California has, of course, its special disadvantages. The beneficence of the climate, as Dr. Ralph E. Smith has observed, is as favorable to insects and diseases as to plants. Californians may take some solace in the fact, however, that scarcely any of the several hundred major pests that constantly besiege its agriculture are indigenous. It has been the constant and extravagant introduction of new crops, new varieties, and new species from all over the world that has made of California something of a pest-and-insect menagerie. Heroic battles have been fought in the state against such enemies as the grape phylloxera, San Jose scale, woolly apple aphid, the coddling moth, the cottony cushion scale, red scale, the pear slug, the citrus mealybug, purple scale, and Hessian fly. A vast amount of money, public and private, has been spent and is spent every year in an effort to fight off the invasions of alien insects and pests. At the moment the most important mystery in California has to do with the shrinking in size of the Valencia orange. For the last five years, experts have sought in vain for an answer to this mystery but the answer still eludes them. "Quick decline," a new citrus virus discovered in 1939, also has them puzzled, as does the "grape leaf skeletonizer," still another new menace. With this background in mind, it is not surprising that the first division of plant pathology in any educational institution in America should have been established at the University of California in 1903. Just as California is a great livestock and agricultural laboratory for the study of plant and animal life, and oil problems, so it is also the world's finest laboratory for the study of plant pathology. It is not by chance, therefore, that the state's plant pathologists have long been recognized as America's leading experts in this field.

Delighted with the unfamiliar climate, the early settlers planted so many different varieties of shrubs and flowers that California gardens began to take on strange and bizarre aspects. A flower show in San Francisco in 1854 brought forth the most amazingly exotic exhibits, scarcely one of which had ever been seen in the state six years previously. The very fact that almost anything could be grown encouraged people to import plants and shrubs from all over the world. From around the rim of the Pacific came every variety of exotic plant and flower. The effect produced by this wild eclecticism was such that one expert complained in 1868 that the campus of the University of California was "more Australian than Californian."

In testing out this novel environment, accidental discoveries have played a role along with conscious experimentations. The discovery that cantaloupes would grow in Imperial Valley was more or less accidental. The discoverer, Duncan Campbell, happened in 1900 to see a few cantaloupes growing in a garden in Indio. He shipped a few boxes to Chicago where the merchants were so impressed with their quality that orders were immediately placed for more. Today cantaloupes and other melons to the value of nearly $10,000,000 a year are produced in Imperial Valley. Until about 1931, it was assumed that winter tomatoes, even in California, had to be grown in hothouses. But, in that year, a man discovered tomatoes growing along a ditch-bank in the Imperial Valley in winter. He began to experiment with the use of "hot caps" and "newspaper teepees" to protect tomato plants and now Imperial Valley has a flourishing winter tomato industry. Out of the sheer delight that people got from planting flowers in California has come a host of industries. The production of flower seed, for example, is a major industry, with California now supplying most of the flower seed of the world. Santa Maria produces $350,000 annually in flower seeds, and nearby Lompoc has a flower-seed industry that has an annual value of about $250,000. The cut-flower industry, based on both greenhouse and outdoor production, is of enormous value. In 1940 California shipped 1,478 express refrigerator carloads of cut flowers and the Japanese florists alone reported gross sales in excess of $3,800,000.

THE ADVANTAGES OF ORGANIZATION

Perhaps the most curious, and certainly the most baffling, aspect of California agriculture is the manner in which a variety of factors have had a mutually re-enforcing effect to produce the peculiar structure of the state's agricultural economy. Distance from markets made for specialization and, conversely, specialization made possible the close organization which made long-distance shipments feasible. Select any one factor which has entered into the development of the state's agriculture and, on examination, it will prove to have a cause-and-effect relationship to every other factor. For example, the real significance of the various factors previously discussed in this chapter cannot be understood apart from the high degree of organization which has long existed and this organization, in turn, has influenced each of the factors mentioned. The fact that so many California products are perishable, has been an important factor in bringing about the high degree of organization. For with perishable products there is no room for argument; if cooperation is necessary, one cooperates, as there is no alternative. The same compulsion exists, of course, in reference to the development of water resources and irrigation systems.

No segment of American agriculture is so highly organized as the California segment. The history of some of the various 400 fruit-and-nut selling organizations dates back to the sixties and fifties of the last century. Area by area, crop by crop, California agriculture is tightly and efficiently organized, with the exception, of course, of the farm workers. Specialization makes for rigid standards and standards imply organization and controls. California not only pioneered in the development of cooperative marketing organizations and worked out the various organizational and legal forms, but the standardization and national advertising of farm products was first extensively developed in that state. The standardization of quality and size of California fruits has now reached the point, according to Dr. Robert W. Hodgson, where it approaches that of modern manufactured products. Starting with the formation of mutual water companies, cooperative organization has expanded to include irrigation, cooperative marketing, pest control, and subsidized research. The marketing of nearly every California crop is today organized, that is, controlled, and with highly profitable results. Although Florida grows a third more oranges than California, its total annual income from this source is about half the California income. In almost every field, organization has been dictated by the nature of the crop and of the environment. Many of the state's specialty crops have highly erratic production records; for example, 306,000 tons of apricots were produced in 1946, 165,000 tons in 1947. Because of the lushness with which crops grow in California, even a slight upward movement in the price for a particular product can quickly produce a glut (as witness the fantastic overproduction of potatoes in 1947 and 1948). To cope with situations of this kind, organization has been absolutely essential. Geographers have pointed out that it was the fact that so many farmers in the Fresno area began to plant grapes that made possible the early cooperative development of an adequate water supply; similarity of interest made for cooperative action. Conversely, the existence of an available water supply was a prime reason for the expansion of the acreage in grapes. The high degree of organization, at every level of California agriculture, has in turn invested the interests that control the state's agriculture with an extraordinary political power which has forced myriad concessions from the state legislature and from Congress.

Perhaps the best summary of the peculiar character of California's agriculture has been provided by Dr. S. V. Wantrup of the University of California when he writes that "climate, soils, location and history have put farming in California in a class by itself. Types of production, yields, and organization differ from those in other states." From whatever point of view one elects to examine the mosaic of the state's agriculture, the conclusion is invariably the same: California's agriculture is "in a class by itself." As in other fields, however, a variety of factors have combined to lessen the advantages which California has long enjoyed. For one thing, other farming areas have begun to utilize techniques of production and methods of organization first developed in California. "California now does not have a corner on new production methods," comments a writer.[1] The advantages which the California poultry industry has long enjoyed have been becoming less significant as other egg-producing areas have successfully imitated California methods and techniques.

The ever-increasing importance of federal marketing controls, also, has tended to level off certain advantages which California has long enjoyed. California was the first state to reap the advantages of large-scale cooperative marketing but, nowadays, many of its crops, such as the truck crops, fruit, nut and specialty crops generally, are excluded from the protection of federal price support programs. One of the reasons for the exclusion, of course, is that these crops are *too* specialized to warrant federal support. In one field after another, California has been losing the advantages which formerly accrued from priority of organization and specialization, as the federal government has expanded its market-control and price-support programs. Hence the sharp increase in such crops as cotton and potatoes in California, crops that do receive price support.

Still another factor is tending to minimize specialty production in California, namely, the sharp increase in population. In the past, California went in for specialty crops to offset the disadvantage of remoteness from markets; but, nowadays, with urban growth absorbing more and more farm land for industrial purposes and with the resident population soaring to new heights, the production of agricultural staples for the local market is becoming of greater importance. The tendency will be, in other words, to shift from specialty to general production. But these shifts-in-use are extremely difficult to bring off in California and it is at precisely this point that its agriculture is vulnerable. With certain crops, notably citrus, avocados, and deciduous fruits, unfavorable returns may force the

[1] Fresno *Bee*, January 25, 1948.

farmer to sell his property; but production will continue under some other ownership long after economic considerations will have called for a shift to some other type of production. Knowing its fabulous cost, a grower will hesitate a long time before he rips out an orange grove. In other words, any major shift in use is, in California, likely to be seriously delayed, painful, and uneconomic.

The temptation to make lush profits from specialty crops also has a tendency, more or less constant, to keep California's agricultural income out of balance. It is very easy, with certain of these crops, to get an enormous over-production; even a slight price increase will often glut the market. Besides, with many of these crops, there is a tendency for production to fluctuate enormously apart from market conditions. A single crop may be raised for three or four different purposes, so it often becomes extremely difficult to reconcile the interests of growers who, although they raise the same crop, have different markets in mind. The divergence of interest between large and small growers has become most acute in many areas and with many crops and now threatens the existence of organizational structures, market setups, and the whole system on which production has been based in the past. In the last analysis, the entire structure of California's peculiar agriculture rests on an extremely shaky foundation, namely, the existence of a pool of mobile, unorganized farm labor. Should the supply of farm labor ever be organized, one can expect that certain advantages which California has long enjoyed would quickly disappear. Many of these advantages have been the by-product of rapid growth in relation to the rest of the country; in other words, maturity has its disadvantages. Here, as in so many other fields, one can see long-term trends at work which are undermining the "exceptionalism," the isolation of California. Nowadays price and cost factors in agricultural production are nation-wide and world-wide in origin, and these are the determining factors. The margin of advantage which California agriculture has long enjoyed is definitely shrinking.

[8]

CALIFORNIA LABOR: TOTAL ENGAGEMENT

❖❖❖

THE CALIFORNIA labor movement has long occupied an altogether exceptional niche in the history of American labor. San Francisco, it has been said, is one of the best laboratories in the nation for the study of industrial relations. Developments have taken place here in a fortnight of history that in other cities have been spread over several decades. The California labor movement, to a degree that is not generally appreciated, has had an important influence on national labor trends. San Francisco was the first major seaport in the world to be thoroughly organized, and it was in this port that the first permanent sailors' union was formed. As the labor capital of the West, San Francisco sent organizers throughout the entire region west of the Rockies and furnished the funds which were used in many western organizing drives. It has been the *total engagement* of labor in California that has, from the beginning, given the California labor movement its distinctive character. The labor struggle in the state has not been partial and limited but total and indivisible; all of labor pitted against all of capital. From time to time, in fact, at fairly regular intervals, California has been convulsed by violent labor struggles. The repetition of this pattern of violence indicates the existence of underlying dynamics of a most exceptional nature. It is the purpose of this chapter, first, to point out the respects in which the California labor movement is exceptional; and, second, to give an account of the how and the why of these deviations.

[*127*]

"No Affinity with Bondage"

The most striking characteristic of the labor movement in California is its deep-rooted and indigenous character. Unions are as old as the state itself. "One is tempted to believe," wrote Lucille Eaves, "that the craftsmen met each other on the way to California and agreed to unite." The mining camps were, in effect, embryo unions which regulated working conditions and prevented unfair competition. The first strike in California took place in the winter of 1849 when the carpenters and joiners of San Francisco struck in support of a wage demand for $16 a day (the prevailing rate was $10). Within a week, the strike was settled on the basis of a compromise of $13 a day which was shortly upped to $14. Sailors first struck in the new seaport of San Francisco in 1850, and the shore workers began to organize in 1853. San Francisco had strong unions before the eastern labor leaders were even aware of the fact that a labor movement existed in this remote outpost of the American frontier. Not only did unions spring into existence overnight —they were born with the founding of San Francisco—but these unions remained local, unaffiliated groups until as late as 1886. No one organized San Francisco; it organized itself.

There seemed to be something in the air, in the social atmosphere of San Francisco, that prompted workingmen to organize. Historians have noted, for example, that the sailors of the port of San Francisco were always "more articulate" than sailors in other American ports. When a slave-owner tried to return Archy Lee, a young Negro, to Mississippi in 1853, the miners of the state, with one voice, prevented the removal of the former slave. Andrew Furuseth, a great California labor leader, once said that the "sea has no affinity with bondage" and so one might say that California has no affinity with any form of bondage. For there has always been some special elixir about California that has prompted men to assert their rights.

From the beginning, also, labor has always been politically orientated in California. A mechanic's lien law was passed in 1850 and a ten hour day statute was enacted in 1853. As early as the

1860's California labor was showing an active interest in politics. In 1877 the Workingmen's Party elected numerous local and state officials; exerted a dominant influence in the adoption of a new state constitution in 1879; and, for a few years, made political history in California. The rise of this new political party represents, as William M. Camp has observed, "the nearest thing to a workers' revolution the West has ever seen." For the first time labor had played a dominant role in the political affairs of a western state. Later, at the turn of the century, the Union Labor Party dominated San Francisco politics for a decade. This early political involvement of labor in California is merely one of many manifestations of the "total engagement" of labor. In no other state has labor been so continuously involved in political action, and from such an early date.

Closely related to this characteristic is the fact, noted by Camp, that "vehement radicalism has marked almost every stage of the growth of the labor movement in San Francisco." Elsewhere radicalism was a late growth in the labor movement; in California it was born, so to speak, with the labor movement. In the 1880's the International Workingmen's Association, a Socialist organization, played a key role in the labor movement; in the period from 1905 to 1920 the Industrial Workers of the World played a similar role; and, in the period from 1920 to 1940, a somewhat similar role was played by the Communist Party. One should note, also, the role which the Socialist Party played in the development of the labor movement in Los Angeles which, for a decade or more, had one of the strongest municipal socialist movements of any American city. This more or less indigenous radicalism which has always gone hand-in-hand with the labor movement is still another indication of the "total engagement" of labor in California.

Another characteristic of the labor movement in the state is to be found in the early and continuous emphasis on joint action. The first central trades assembly was formed in San Francisco in 1863; the first statewide federation of labor in 1867; and the first effort to unite the waterfront unions took place in 1886. The tendency of labor to federate in California has paralleled a similar tendency on the part of employers to unite. Some of the first employer organizations in the nation were formed in San Francisco, and as early as 1888, one finds unions being pitted against employers as a group. Industry-wide collective bargaining, in fact, had its genesis in California. The history of labor relations in California, as the La Follette Committee discovered, is essentially a history of the struggle between "associations of employers" and "federations of unions." "To a greater degree than this Committee has found elsewhere," reads the report, "associations of employers in California have played a leading role in fixing labor policies, and have been able to impose their influence upon the social and economic structure of the state." In short, the history of labor in California is really not a history of the struggle of unions to achieve recognition but of a struggle for power between organized labor and organized capital. From the outset, both sides have been fully engaged, totally committed. The nature of this engagement accounts for the periodic convulsions in the state's social history in which periods of intense conflict have alternated with periods when labor's resentment smoldered beneath an apparently tranquil surface. Both the scale and bitterness of the labor struggle in California are most remarkable when one realizes that California did not become a major industrial center until well after the turn of the century.

A final characteristic of the labor movement in California is to be found in the fact that, at various periods, labor has spoken for large masses of people in the state who were not functionally a part of the labor movement. For many years, the California labor movement also included within its ranks a large petty bourgeois element. In fact it has only been of recent years that industrial workers, as such, have come to be the mainstay of the labor movement. Obviously special influences have shaped the labor movement in the state; otherwise it would be difficult to account for the paradox of a strong labor movement in a non-industrial state.

This brief specification will suffice to make the point that there has always been "something peculiar" and different about the labor movement in California. As with other aspects of the state, the key to an understanding of California's peculiar labor dynamics is to be discovered by concentrating attention upon the exceptional qualities of the state itself, the things that make it different.

California *Labor: Total Engagement*

"The Magic Scepter"

That a strong labor movement should have arisen in early-day San Francisco is in part to be explained by the key location of the city. Here was a centrally located harbor on a coast where, as Miss Eaves points out, "the mountains crowd close to the oceanside and where but few indentations permit a safe entrance for commerce." Until the completion of the transcontinental railroad, San Francisco was the key point of entrance and exit from the state. With the Sacramento and San Joaquin Rivers entering into the bay, San Francisco was in a position to control the commerce between the outside world and the gold camps. It was to San Francisco that the rich came to spend their money; that the unemployed came in search of new opportunities; that the discouraged came to seek exit from the state. This city assumed a political power comparable to its social and economic power. Over a period of many years, control of San Francisco, more particularly control of the waterfront, was tantamount to control of the entire state. The unique geographical position which San Francisco occupied in relation to the rest of the state, and to the entire West, gave labor its great opportunity in California.

What invested the centrally located position of San Francisco with such extraordinary significance, however, was the discovery of gold. Gold, in relation to labor, was indeed "the magic scepter." For reasons already pointed out, the California mining frontier was unlike other mining frontiers because of the extraordinary democracy of opportunity which prevailed. "Nowhere in the world," writes Miss Eaves, "has there been a more favorable economic environment, nor more freedom for social and political experiments than in California." It is significant that the first labor legislation adopted in the state was an ordinance of 1847 aimed at preventing the desertion of ships by sailors. Men simply could not be kept at jobs other than mining. "Desertion," in fact, was the major labor problem—desertion from ships, mills, farms, stores, foundries. "In the days of '48 and '49," writes Dr. Ira Cross, "the employer as such was virtually unknown." The shortage of labor and the wealth of economic opportunities which existed created an extraordinary, and never fully recaptured, opportunity for labor. The circumstances suggested organization. No precedents were needed; no external stimulus was required.

Social factors, born of the same situation, re-enforced the economic factors. "Every man was a laborer," writes Dr. Cross, "whether or not he had previously been a teacher, lawyer, mechanic, preacher or sailor. Physical labor was honorable. Class lines and class distinctions were forgotten, and a universal spirit of rough democracy prevailed. This wholehearted democratic spirit of the mining days permeated virtually every phase of early California life." To illustrate the universality of this spirit of labor, suffice it to say, that in a strike of thirty carpenters in 1849 it was discovered that three of the strikers were preachers, two lawyers, three physicians, six bookkeepers, two blacksmiths, and one was a shoemaker. Long after the gold rush had vanished, the tradition of high wages, of the honor and dignity of labor, continued to create a congenial social milieu for trade union activity. Over a period of many years, as Miss Eaves has noted, "the workmen as a rule had the sympathy of the public."

The geographic isolation of California also strengthened labor's position. In the crucial decades prior to the completion of the Central Pacific, it was quite impossible to recruit strikebreakers, or to flood the labor market with new recruits. Distance threw a protective tariff, so to speak, about the local labor market. No picket line could have been more effective than the distance which separated San Francisco from the centers of population. When the bakers of the city struck in 1863, the employers had to send to Hamburg for strikebreakers and by the time they arrived the strike had been settled. Not only did workers have a magnificent opportunity to organize but they had the ability to enforce their demands. The factor of distance operated in still another way for it invested the local unions with an almost complete freedom of action. Even if the San Francisco unions had been affiliated with national organizations, it would have been quite impossible for the parent union to have imposed its discipline upon them. "So isolated was the city itself from Eastern centers of labor," writes Camp, "that the strongest point in favor of solid labor unionism was its independence." Hence the strong tradition of local autonomy which has long prevailed in the labor movement in California. By and large, local autonomy makes for strong unions that stand on their own feet and fight their own battles.

Labor's opportunity in California was, of course, capital's special disability. From the outset employer groups felt compelled to experiment with strong-arm tactics in order to offset the advantage which labor possessed. The tradition which sanctions the use of extra-legal tactics by employer groups is almost as old as the labor movement in California. The difficulty which shipowners faced in maintaining crews accounts for the fact that San Francisco was the first major port to permit unrestricted crimping, i.e., the procurement of sailors by decoy, fraud, and violence. In fact the word "shanghai" originated in San Francisco. Crimping was sanctioned from the earliest time by both the ship-owners and the municipal authorities. The system existed for so long that it came to be regarded as part of the business of shipping itself and no more to be questioned than one would question the accuracy of a nautical chart.

Since capital was for so many years at a distinct disadvantage in its dealings with labor, a tradition of violent tactics arose which, of course, had an enormously stimulating effect on labor organization. Each side was driven to take strong measures against the other: labor to exploit its extraordinary opportunity; capital to cope with an exceptionally powerful labor movement. It is this peculiar relation between labor and capital which Camp had in mind when he wrote that "just as San Francisco was the first major port to permit unrestricted crimping, so also was San Francisco to become the first airtight 'labor town.'" The tradition of strong-arm employer tactics also accounts for the vehement radicalism which has gone hand-in-hand with labor organization in California. The "direct action" of the Wobblies was the counterpart of the "direct action" of the employer groups.

The wonderful opportunity which labor possessed during the gold rush period was not, moreover, something that once existed and then was lost; to a considerable extent it has continued to exist. San Francisco was not only the first boom town in the West but the one town that continued to boom. Although the bonanza days soon passed, nevertheless the rapidity with which California continued to grow created a most favorable economic environment for labor organization. Not only was the growth of the state phenomenal, but it recovered more rapidly than other areas from periods of depression. In a study of business cycles in California, Dr. Frank L. Kidner has found that "there is an apparent tendency for economic activity in California to recover from a business depression more rapidly and more fully than is true of the United States as a whole." In the booms which invariably followed the periods of depression, labor possessed marked advantages in relation to capital and it never hesitated to exploit these advantages. This recuperative power, the ability to bounce back quickly from depressions, reflected the continued migration of population to California and the fact that the state remained a land of new and expanding business opportunities. It has been this phenomenon of "quick recovery" which explains the fact that trade unionism has flourished in California during periods when labor in other sections of the country has been caught in the backwash of the economic cycle. Boom times, as organizers know, are good times to organize and the history of California is a history of booms.

Another secret of labor's power in California consists in the selective force of migration. A large part of the skilled labor force of San Francisco was made up of foreign-born workers who brought a knowledge of trade-union organization to California. For example, there were unions of German-speaking cigar-makers, brewers, bakers, and cabinet-makers. The Sailors' Union, which served as a training school for trade unionists in San Francisco, was largely made up of men who were natives of Sweden, Norway, and Finland. Between 1889 and 1903, 13,796 men left this organization to enter other occupations; half of them were natives of these three countries. "In a society where all were strangers," writes Miss Eaves, "the possession of a common trade would furnish the most natural and promptly recognized bond of union." And this bond, of course, was strongest where it happened to be identical with the bond of a similar language and cultural background.

The selective force of migration, however, operated in still another way. The presence of a large number of Chinese restricted to undesirable jobs had the effect of discouraging the migration of unskilled workers. The fact that the types of jobs most difficult to organize fell to the Chinese made it all the more easy to organize the skilled trades. Ordinarily the existence of a large pool of unskilled labor operates as a threat to the standards which labor seeks to establish in the skilled trades; but, in this case, the unskilled were racially distinguishable and were under a great handicap by reason of the language barrier and other factors. It was extremely difficult, therefore, to recruit apprentices from this group, a fact which served to invest the skilled trades with a special degree of protection. The absence from the labor market, also, of women and children tended to protect the standards which organized labor had established.

Still another factor underlying labor's exceptional opportunity in California is to be found, as Miss Eaves noted, "in an entire absence of that conservatism that comes with the more gradual accumulation of wealth." California has always been a rich state and richness makes poverty anachronistic. The quickness with which fortunes were amassed had bred in California a remarkable fondness for luxury which was ostentatiously exhibited. America has surely produced few millionaires who were less inhibited than the millionaires of early San Francisco. The circumstance that everyone knew that much of this wealth had been won by sheer luck created a disposition to demand a cut, to insist on high wages. Once the tradition of high wages was established, the ex-miners were psychologically unprepared and unwilling to accept a return to "normal" wage rates. As one historian has pointed out, they insisted "on the wages to which they had become accustomed." At this point the recurrence of booms becomes an important factor. Experienced Californians know that the state's booms do not last forever and that they must be quickly exploited. Hence every boom has touched off a hot labor-capital conflict.

Lastly it should be noted that the rise of San Francisco to world importance as a seaport occurred during the most formative years of the world labor movement. The year 1848 is of crucial importance in the history of European and American labor, and 1848 is the natal year for California.

THE ECONOMICS OF EXTREMISM

Labor unions in California have been compelled by the nature of the economy of the state to federate, to seek alliances, and to found assemblies. The second largest state in the union, California is a highly developed economic area. All forms of economic activity are embraced within its borders. Its high level of cash farm income; the value of its mineral and forest products; its fisheries and oil fields; its canning and processing industries—these and other factors have served to make it an economic empire in itself. Since it is not one thing economically, but many things, labor has been compelled to reach out, to expand the area of organization, and to consolidate its gains. The labor market is as large, as interrelated and as interdependent as the state's economic activities. In such an economic area, nothing less than complete organization can possibly safeguard the interests of labor.

In California, also, agriculture has a unique relation to industry which arises from the dependence of agriculture on the export market and the accompanying reliance upon the processing, handling, and transportation industries. The very nature of many California crops brings a host of industries into close and intimate relation with agriculture. In 1948, some 6,352 trucks were used in transporting produce from the Imperial Valley alone, which suggests the reliance of agriculture upon transportation. In fact, it is often difficult to classify a particular industry in the state as being primarily "agricultural" or "industrial." The type of labor used in many agricultural operations shades off imperceptibly into the type of labor used in the handling and processing industries. The interdependence of so many industries has naturally encouraged labor to achieve, if possible, total organization.

For many years the economic life of the state has been dominated by two urban centers, San Francisco and Los Angeles, and the existence of these two competing centers has tended to divide the state into two major economic areas. The fact that large agricultural and tributary areas have been so highly dependent upon two major urban areas has given the urban areas an enormous power to influence labor relations in their respective hinterlands. The determination of labor policies in these two urban areas has affected labor policies throughout their respective regions; conversely, control of labor policies in the tributary areas is vital to control of labor policies in the urban centers. If the two major urban centers had not been highly competitive, each might have been able to ignore or to tolerate conditions in the hinterland areas which were inconsistent with or tended to undermine urban labor policies; but they have never been able to afford this tolerance.

The rivalry between the two centers, moreover, has always been accentuated by the fact that San Francisco was vitally dependent upon its port; whereas Los Angeles was late in developing a manmade port. The close relation between the Port of San Francisco and California agriculture can be shown by reference to the principal commodities which were exported through the port in 1938. Agricultural products totaled $75,744,046; other extractive products, such as wood, bulk oil, etc., totaled $44,276,415; and other products $19,599,524. So far as the economic activity of the northern and central portion of the state is concerned, the Port of San Francisco is the bottleneck. On more than one occasion, therefore, control of the San Francisco waterfront has carried with it, as a rich prize, indirect control over a large part of the economy of the state. Just as the waterfront has been the scene of innumerable labor struggles, so San Francisco labor subsidized the fight to organize Los Angeles, for to the extent that commerce and industry shifted to Los Angeles the advantage of waterfront control was weakened. Labor had to expand, therefore, in order to protect what it had achieved.

A large part of the California labor market has always been seasonal in character. Employment expands and contracts in the canning, processing, and handling industries as it expands and contracts in agriculture. The casual nature of waterfront employment invests it with some of the characteristics of a seasonal labor market. Faced with this situation, unions have been compelled to extend their control over the entire labor market; in fact the fight to control the *supply* of labor has been, perhaps, more important than the struggle to raise wages or to improve working conditions. For precisely the same reason, employers have strenuously resisted every attempt by labor to control the entire labor market. With seasonal employment being of such crucial interest, it is extremely important, from the employer's point of view, that the labor market should be kept unorganized and fluid. Seasonal industries in California cannot tolerate any interruption in work schedules. Peaches must be picked at a certain time; they must be processed on schedule; and they must be shipped on time to reach distant markets.

In short, the nature of the state's economy has always catapulted labor and capital into an intense struggle for control of the labor market. Since the labor market is as diverse as the products produced, both sides have sought to gain strength by combination. They have reached out, also, for allies and have constantly sought to enlist the public on their side. It has been the compulsion to reach out and control *related* lines of economic activity that accounts for the continued emphasis which the California labor movement has always placed on such weapons as the secondary boycott, the sympathetic strike, and "hot cargo" tactics. To keep the Chinese relegated to the unskilled trades, California labor made the first extensive use of the boycott in this country. The union label, which has long since become part of labor's arsenal of weapons, was first used in California. These various weapons, the boycott, the union label, the refusal to handle "hot cargo" and so forth, have been of great importance in a state whose economy was so interrelated and interdependent as that of California's.

The same compulsions have driven both labor and capital in California to achieve, in their respective fields, total integration. "Labor unions and employee groups," reads the report of the La Follette Committee, "have been driven to cooperate with one another to a greater degree, perhaps, than in any other section of the nation." The same, of course, is true of the employer groups.

The Ship Owners' Protective Association of the Pacific Coast, formed in San Francisco in 1886, was the first association among employers to be formed in this country for the exclusive purpose of dealing with labor. Not only was industry quick to use the collective approach to labor problems in California, but employer associations are almost as old as the trade union movement itself. California has always had a pattern of organized anti-unionism. Industrial employers in the state have been more solidly arrayed, as a class, against labor than in any other state. To such an extent has this been true that, with the formation of the Board of Manufacturers and Employers in San Francisco in 1893, the day of the independent, isolated businessman in labor relations was gone. From 1900 to 1940, the Merchants and Manufacturers' Association in Los Angeles coerced the small industrialist and businessman into following the labor policies which its directors had decided upon. In no respect is this control-by-association more strikingly illustrated than in the "license system" which the Industrial Association of San Francisco used to wreck the building trades unions in 1921. Under this system, every contractor had to agree in writing to operate an open shop before the material dealers would furnish him with materials and supplies. Confronted with this type of united opposition, unions have been compelled to resort to extraordinary tactics in order to survive.

LABOR'S CURIOUS DUCKLINGS

The major paradox about the labor movement in California consists in the fact that a powerful labor movement should ever have arisen in a state which, prior to 1900, was largely non-industrial. Yet by 1900 San Francisco was recognized as not only the most tightly organized city in America but as the stronghold of trade unionism in the United States. Obviously the labor movement in California must have included elements which are not ordinarily thought of as part of labor. In California there were three such elements: the small shopkeeping element; a large section of the rural population; and a sizable element of what today would be called "white collar" workers. How was it that these elements became allies and, in some cases, integral parts of the labor movement?

The answer is to be found in Camp's statement that the fear of Chinese competition in California "brought about the rise of such a great wave of emotional class consciousness that it swept obscure opportunists into public office." But it did more than sweep opportunists into office; it drove thousands of shopowners, farmers, and clerical workers into the camp of organized labor. If the gold rush had not brought a tidal wave of white settlers to California, it is altogether possible that the whites might have formed a tightly knit plantation-like economy based on the use of Chinese labor; but the whites were too numerous in relation to the Chinese to form a ruling clique. The alternative was to organize and thereby force the Chinese into the undesirable positions. A better alternative, of course, would have been to organize the Chinese also, but the language and cultural barriers were too great to make this a feasible alternative. It was, in any case, the threat of competition from Asiatic labor that made for solidarity and invested labor in California with a political power far stronger than it has ever possessed in any other state.

The potency of anti-Chinese agitation as "an emotional class consciousness" consisted in the fact that it tended to fuse with class lines. At an early date, J. Ross Browne reported that he could find "among the influential and respectable class" little antagonism to the Chinese. "The objections against them," he said, "are purely of a local and political character and come from the lower classes of Irish." By and large, the upper classes consistently favored unrestricted immigration; the lower classes as consistently opposed it. By utilizing this unity of feeling against the Chinese, labor was able to build the most powerful alliances. Nor was anti-Oriental agitation a passing phase in California politics; in various phases it persisted for seventy years or longer. It goes without saying, of course, that this movement had some extremely ugly implications; but it was certainly the force that held labor together.

After 1900 anti-Japanese agitation was used for the same purpose as anti-Chinese agitation had been used over a period of thirty years, i.e. to build a powerful labor movement. From 1900 to

1910 a union charter in California was, in some respects, primarily significant as an authorization to engage in anti-Japanese agitation. The Japanese represented a more potent threat to the lower middle class and middle class than the Chinese for they demonstrated a remarkable ability to move up into the self-employed and farm-owner category. The threat of this competition, real or imagined, drove thousands of people into labor's ranks not only in the cities but in the small towns and rural areas. After 1900, as one labor journal put it, the unions experienced a "Pentecost breeze." In fact it is doubtful if any state ever felt the ardor for organization that then prevailed in California. All sorts of occupations and callings were organized and charters were "signed for and hung in meeting houses until they covered the four walls." But, as this same labor journal pointed out, "very few of these unions were trade unions. . . . The labor council gathered under its wings a most varied collection of eggs and hatched some curious ducklings and labeled them trade unions." As one reads through lists of unions formed during this period one notices butchers, barbers, bakers, picture frame makers, cloak-makers, tailors, milk wagon drivers, art glass blowers, blacksmiths, and many similar occupations which usually fall into the "little business" category.

Ordinarily there is no more inveterate if misguided opponent of organized labor than the small shopkeeper. It is the history of small shopkeepers that they are usually more capitalistic than the capitalists. They are also notoriously chauvinistic; in fact it was their tendency toward chauvinism that brought them into the California labor movement in droves. Many of these elements, of course, were never thoroughly integrated with the labor movement and they began to drop out as the anti-Oriental agitation passed out of the control of the labor leaders. These were the elements that kept some of California's most corrupt "labor" politicians in power for many years, thereby bringing great discredit to the labor movement.

Regardless of the price that labor ultimately paid for its espousal of the anti-Oriental movement, there can be little doubt that this movement, from an opportunistic point of view, paid great dividends to labor. In 1911, 39 out of 49 labor measures placed before the state legislature were adopted and a similar record was made in 1913 and 1915, with the result, as Dr. Cross has pointed out, "that California took a prominent place among states interested in conserving the welfare of the working class"—that is, the non-Oriental working class.

THE PATTERN OF VIOLENCE

Since labor was totally engaged with capital from the earliest date, it is not surprising that the history of labor in California should be a history of labor's strenuous and often violent thrusts for power, and of the equally violent counter-repression invoked by capital. It has been this periodic outbreak of class warfare on a large scale which has been so largely responsible for the continued political instability of the state. Even in those periods in which labor has held the upper hand, fear of the expected and inevitable counter-attack from the organized anti-union forces has driven labor to seek still further power. "Cease fire" orders have been given from time to time but until the federal government began to intervene in labor relations there was no real peace between capital and labor.

Without going into the full details, it can be said that there have been four major labor-capital battles in California. The first, which occurred in the period from 1886 to 1893, had its genesis in a determined effort on the part of the employers to break the power of the unions. In 1886 there had been a serious waterfront strike, which is generally taken to mark the beginning of San Francisco's famous waterfront warfare, and an important strike by the brewery workers. In both cases, the contest had quickly developed into a fight between *groups* of unions and *groups* of employers. The employers were particularly disturbed by the formation in 1891 of the Coast Seamen's Union, a truly remarkable labor organization and the first stable organization of its kind to be formed in the world. Embracing the entire Pacific Coast, the union was centrally directed from San Francisco with agents in every west coast port. Wherever the coasting sailor went, into whatever port, his membership card was recognized and he enjoyed the same protection as every other sailor in that port.

There had also occurred, in 1890, a bitter fight between the Iron Trades Council, a federation of metal workers, and an employer organization known as the Engineers' and Foundrymen's Association. In each case, an issue had been fought out between a group of unions on the one hand and a particular employers' association on the other. The employers, therefore, decided to form an all-inclusive employers group—the Board of Manufacturers and Employers formed in 1891—and to break up, if possible, the combinations of unions that had developed. This particular struggle culminated in a second waterfront strike in 1893 which labor lost largely because the explosion of a bomb on Christmas Day in front of a non-union boarding house, killing eight men and wounding many others, alienated public support. As a consequence of this defeat, the unions of San Francisco were, for the time being, largely destroyed or at least demoralized to the point where little unity or strength remained. This first battle, therefore, resulted in an unqualified victory for capital.

But, by the turn of the century, California was again booming. The Spanish-American War, the annexation of Hawaii, the gold rush to Alaska, and other factors stimulated a flurry of industrial activity in the state. Both sides, of course, immediately began to prepare for a resumption of the earlier battle. In this case the unions took the offensive since they feared that the employers were plotting another systematic campaign against them. In a great organizing campaign the number of union members was doubled in a year. The State Federation of Labor was formed in 1901, and the City Front Federation, a loosely knit federation representing some 13,000 waterfront workers, came into being the same year. The employers promptly formed an all-inclusive employers' group, the Employers Council, and proceeded to raise a war chest of $250,000 which was precisely the amount the City Front Federation had in its treasury.

The second great struggle began on July 30, 1901, when the waterfront workers struck, but the situation quickly developed into a tangle of sympathetic strikes as the two great contending forces moved into action. For three months the harbor was crippled. In the course of this strike, 5 men were killed and 300 assaults were reported. The violence was so great that both sides seem to have exhausted themselves and a mutual cessation of hostilities was finally negotiated without either side having won a clear-cut victory. In effect, however, the unions won this round because they emerged from the battle stronger than when they had entered it. "There is a kind of fighting which makes the enemy stronger," reported Ray Stannard Baker at the time, "and that was the method of the San Francisco Employers' Association. It was an example of how *not* to combat unionism." A few weeks after the strike was called off, the Union Labor Party won a smashing political victory in San Francisco and remained in undisputed control of the city administration for a decade. In the wake of this strike, in fact, San Francisco emerged as the first "closed shop" city in America.

The third great battle developed shortly after the outbreak of the first World War. The war, of course, immediately brought about a sharp increase in the volume of cargo moving through the port and both sides promptly squared away for another slugging match. In 1916 the longshoremen went on strike, bottling up some $2,500,000 in exports. As in the prior struggle, the farming and business interests of the hinterland demanded that the San Francisco employers' group should break the strike. The murder of a striking longshoreman on June 21 seemed, for a few days, to tip the scales of public opinion in favor of the unions. But, while the strike was still on, the tragic Preparedness Day bombing took place (on July 22nd) in which some 10 people were killed and 40 seriously injured. Out of this fateful event, of course, came the infamous frame-up of Tom Mooney. The bombing threw the weight of public opinion against the unions, the strike was lost, and, at the height of the excitement, the city adopted an anti-picketing ordinance by a vote of 73,993 to 68,570.

Following its earlier victory in 1901, organized labor in San Francisco had decided that the time had come to organize Los Angeles, "the open shop citadel of America." Just how important this organizing drive was, in terms of protecting the closed shop in San Francisco, can be shown by the fact that in 1900 San Francisco had 66 per cent of the total organized trade union membership of the state by comparison with 6 per cent in Los Angeles. In 1910, 65 per cent of the trade union strength was in San Francisco and only 8 per cent in Los Angeles. Viewing open shop Los Angeles as a threat to everything it had achieved in San Francisco, the labor movement proceeded to raise nearly $500,000 for an organizing campaign. There the trouble started on May 19, 1910, with a strike of brewery workers, followed by a strike of metal workers and of Mexican workers on the street railway. To break these strikes, the Merchants' and Manufacturers' Association drafted an anti-picketing ordinance which is known in the labor histories as the model for all the anti-picketing laws and ordinances in the country.

Within a few weeks after the adoption of this ordinance on July 16, 1910, over 470 workers had been arrested; but, almost as fast as they were arrested, Los Angeles juries acquitted them. This particular struggle culminated in the dynamiting of the Los Angeles *Times* on October 1st, 1910, in which 21 men lost their lives. This dreadful explosion, and the plea of guilty which the McNamara brothers entered a year later, set the cause of labor back for at least two decades in Los Angeles. Previously uncompromising in their anti-union attitude, the open shop employers of Los Angeles used this event in a most spectacular and devastating manner to swing community sentiment to their narrow purposes. What the dynamiting of the *Times* was to Los Angeles, the Preparedness Day bombing was to San Francisco: both events symbolized a crushing defeat for the labor movement.

All this while, however, there was another "labor movement" in California spearheaded by an outlaw, revolutionary organization, the Industrial Workers of the World. There was only one delegate from California at the meeting in Chicago on June 27, 1905, at which the I.W.W. was formed but, by 1910, the wobblies had 11 locals in the state and nearly a thousand members. It was Local No. 66, which Frank Little had organized at Fresno, that launched the first of the famous wobbly free speech fights in California. The campaign opened with an outdoor meeting on October 16, 1910 at which Frank Little, one of the speakers, was arrested and given a jail sentence by a jury which he contemptuously referred to as "composed of Bourgeois cockroaches and real estate grafters." In subsequent meetings, first 10, then 15, then 25, and 50 people were arrested and, finally, lawless elements in the community burned the wobbly headquarters. Fire hoses were used by irate police in an unsuccessful effort to keep the arrested wobblies from singing in jail. The campaign was finally settled, six months later, by the appointment of a mediation committee and at least a partial vindication of the right of free speech was secured.

The wobblies, of course, were quite free of the chauvinism which prevailed in the California labor movement at this time. They repeatedly attacked the "yellow peril" agitation and sought, without too much success, to organize Mexican field workers and other minority groups. Although many of the labor leaders of California of this period were of foreign birth, most of the wobbly leaders, ironically, were Old Americans with names like Dunn, Ryan, Olson, Sherman, and Eaton. The wobblies had real influence with the casual and seasonal workers of California, notably the waterfront workers, the lumberjacks, and the field and cannery workers. Their informal organization, the tactic of organizing on the job, the use of quick strikes, and the roving and migratory nature of the organization itself, made the wobblies effective pioneers in the effort to organize seasonal and casual workers.

Following the Fresno free speech fight, San Diego adopted on January 8, 1912 an ordinance limiting the right of free speech. The wobblies promptly moved in and launched a sensational fight to have the ordinance revoked. Although they had not more than 50 members in San Diego, it has been estimated that nearly 5,000 people took part in this campaign. Michael Hoy, a wobbly, was kicked to death in jail and another member, Joe Mikolash, was shot and killed. When jailings failed to break the spirit of the wobblies, a vigilante mob aided by the police rounded up several hundred men, made them "run the gauntlet," beat them with clubs and fire hoses, and drove them out of town. The issue reached such a pitch of excitement that Governor Hiram Johnson sent Harris Weinstock to San Diego to make an official report and investigation. In part because of this excellent, clear-headed report, the wobbly campaign was finally successful and the right of free speech was vindicated.

The wobbly campaign in California came to a climax with the famous Wheatland "hop pickers'" riot of August 3, 1913, in which four people were killed.

Some 2,800 hop pickers, representing a wide diversity of nationalities, had been recruited by ads for work on a ranch owned by one of the largest employers of farm labor in the state. The pickers included a large number of women and children. On arriving at the ranch, the pickers found that the wage rates varied from day to day, depending on the number of pickers on hand, and that the "bonus"—which was advertised—was actually a "hold-back" forfeited if the worker left the job. Widely distributed, the ads had brought in about 1,000 more pickers than were needed. Average daily earnings were found to be about 90¢ or $1. The conditions at the camp may be indicated by the fact that 8 small toilets had been built to accommodate 2,800 people and that there were no separate toilets for women. The riot was touched off when law enforcement officials attempted to break up a protest meeting which a group of wobblies had called on the ranch.

To the wobbly movement, the Wheatland Riot had much the same significance that the Preparedness Day bombing and the dynamiting of the *Times* had for the labor movement; the three events, in fact, were part of a much larger pattern of violence in industrial relations. The Wheatland Riot is of great historic importance for it marked the beginning, in a sense, of intense labor strife in California agriculture. There had been earlier incidents, of course, but this case focused national attention for the first time on the miserable plight of seasonal field workers in California. Out of this incident came the prosecution of Richard Ford and Herman Suhr, both of whom were convicted in one of the most famous "labor trials" in the state's history. Along with Tom Mooney, J. J. McNamara, and J. B. McNamara, "Blackie" Ford and Herman Suhr acquired legendary fame as "labor martyrs." In the context of this chapter, the Wheatland affair is of importance for two reasons: it marked the extension to agriculture of the pattern of "total engagement" which had long characterized labor relations in California; and it emphasized, once again, the manner in which repressive employer tactics consistently precipitated radical protests.

In a broad historical sense, the third chapter of labor violence in California came to its climax with the adoption on April 30, 1919, of the Criminal Syndicalism Law. Although Idaho has the unenviable distinction of having adopted the first statute of this kind, the California act received the most notoriety because it was more widely enforced than any similar legislation. Criminal syndicalism acts in the other states soon became "dead letter" statutes but the California act was systematically enforced. In a five-year period following its adoption, 504 persons were arrested, bail was usually set at $15,000, and 264 of those arrested were actually tried. At least 34 cases, arising under this act, went to the appellate courts. Of those arrested, 164 were convicted and 128 of these were sentenced to San Quentin Prison for terms which ranged from one to 14 years. The emphasis given the enforcement of this act in California is not surprising for its adoption represented the culmination of seventy years of intense anti-union activity on the part of employer groups. It was, in effect, the logical end-product of the "total engagement" between capital and labor in California.

The fourth "engagement" took place in the 1930's and involved, first, a recrudescence of the waterfront warfare which had become more or less endemic in San Francisco, and, second, a series of great strikes in agriculture. Between January 1, 1933, and June 1, 1939, approximately 180 agricultural strikes were reported in California; farm labor strikes were reported, in fact, in 34 of the 58 counties of the state. All in all, some 89,276 workers took part in these strikes for which no parallel of any kind can be found in the history of American labor. Civil and criminal disturbances were reported in 65 of the strikes, with hundreds of arrests, 14 "violent" strikes, several deaths, and considerable property damage. The ferment of these years reached its climax with the "general strike" in San Francisco, July 16th to 19th, 1934, which was called to protest the killing of two waterfront workers on "Bloody Thursday," July 5th. Although the general strike collapsed, the waterfront workers won a great victory which was followed up, one year later, with the formation of the Maritime Federation of the Pacific. As much as anything else, perhaps, it was this upsurge in labor activity, following the suppressions of the period from 1910 to 1924, that brought about the election of Governor Culbert L. Olson in 1938 whose first official act was the issuance of a pardon for Tom Mooney. The fourth round, in short, was won by labor and, with the adoption of the National Labor Relations Act in 1936, the labor movement achieved a new maturity and succeeded, at long last, in breaking the power of the employer organizations and in organizing "open shop" Los Angeles.

This greatly abbreviated statement of the pattern of attack and counter-attack should indicate the all-out character of the labor struggle in California. The unevenness of this struggle, with first labor and then capital, achieving the upper hand, largely accounts for the marked political instability of California. With the enactment of federal legislation assuring labor's right to organize and safeguarding the principle of collective bargaining, some of this political instability has disappeared and there has been a noticeable leveling-off of the sharp peaks and valleys of industrial conflict. But the end is not yet in sight if only for the reason that the processes which have finally brought a measure of peace to industry have still not been applied in agriculture. For California suffers from an ancient, malignant, and festering cancer—its notorious farm labor problem—which it is the purpose of the next chapter to describe.

[9]

CALIFORNIA'S PECULIAR INSTITUTION

◇◆◇

CALIFORNIA'S PROBLEMS are, of course, as exceptional as its advantages. Among these exceptional problems none is of greater importance than the state's seventy-year-old farm labor problem. Other states have a farm labor problem but California's is unique in malignancy, magnitude, and virulence. Farm labor is California's "peculiar institution" in much the same sense that chattel slavery was the South's peculiar institution. Today as yesterday, the farm labor problem is the cancer which lies beneath the beauty, richness, and fertility of the valleys of California. For more than seventy years, a large portion of the state's population has lived in a kind of social "no man's land," often disfranchised, consistently unrepresented, and on many occasions, brutally repressed. The consequences of this ancient denial of the democratic promise of California to a large section of its population have been far reaching. Suppression breeds arrogance as well as rebellion and the marks of arrogance appear wherever farm labor is employed. Over a period of many years, the farm labor problem has become part of the structure of the state's economy; hence its solution presents enormous difficulties. Nowadays the problem is thoroughly understood and the facts are notorious but nothing approximating a solution has been achieved. A knowledge of the nature of this problem is fundamental to an understanding of California.

THE GRAPES OF WRATH ARE STORED

In the wake of severe droughts in 1862 and 1864, the cattle industry of California collapsed and the period of the great bonanza wheat farms began. The availability of land in large parcels, the long dry season, and the evenness of the topography permitted an efficient use of large-scale machinery and equipment, and wheat farms of twenty to forty thousand acres were not uncommon. California was "nearer" to the wheat markets of the world, by reason of its access to Pacific ports, than the other large wheat-raising areas and, for several years, it ranked among the most important wheat-producing states. As a result of the extensive use of machinery, little labor was needed until the harvest period when thousands of workers were required. In this manner, a migratory labor movement came into being which was made up, originally, of Indians, tramps, and, later, of Chinese workers.

In the *Morning Chronicle* of San Francisco for September 5, 1875, one finds the following description of the farm labor problem in California:

> The farm labor problem of California is undoubtedly the worst in the United States. It is bad for the farmers themselves, and worse, if possible, for those whom they employ. In many respects, it is even worse than old-time slavery. That, at least, enabled the planter to know what labor he could depend upon in any emergency, and made the labourer certain at all times of shelter, clothing, food, and fire. Our system does neither. The farmer must take such help as he can get —hunting it up when most hurried and paying whatever is demanded. The labourers themselves, knowing that they cannot be permanently employed, demand high prices, do their work carelessly, and start out on a tramp for another job. Under our system large numbers of men are wanted for a short time; more than any farmhouse can accommodate, even if the employer dare trust so many strangers within his walls or admit them into his family circle.
>
> The result is that labourers are compelled to sleep in barns, outhouses, or in the open fields. Men seem thus to have been thrown outside of social influence, and even if at the outset possessing good impulses and habits, they become, in a short time, desperate, degraded, or criminal, and perhaps all three.

Make allowance for certain changes and here you have an accurate, up-to-the-minute description of the farm labor system of California; there are no basic changes except that the problem is more acute today.

With the completion of the transcontinental railroad and the expansion of irrigation systems, a great increase in orchard acreage took place in California. Little equipment or machinery could be used in the orchard crops so that the demand for labor increased with each new planting. At the outset, this demand was met by the use of Chinese who had become available in large numbers with the completion of the railroad and their exodus from the gold fields. In 1870, Chinese made up one-tenth of the farm labor supply; but in 1880 they constituted one-third and by 1884 half of the farm workers. The existence of a wage differential greatly stimulated the antagonism of other workers toward the Chinese, and resulted in a campaign to drive them from the fields. Anti-Chinese riots were reported in many rural areas and small towns and, as the agitation mounted, the federal government was compelled to exclude further Chinese immigration. After 1884 the Chinese declined in relative importance since many had fled from the state or retreated to the towns and cities. At the same time, a series of depressions drove more and more "white" workers into farm areas.

Thus by the turn of the century a system of farm labor had come into being which was based originally upon the exploitation of seasonal alien labor. Between 1850 and 1860 farm wage rates had declined rapidly and continued to decline until 1900. In fact, wage rates had come to be stabilized at a level which made it impossible for seasonal workers to earn a living annual wage. But, in this same period, the value of California farm land rose substantially and came to be capitalized, ultimately, upon the basis of actual and anticipated profits accruing from the extensive use of cheap labor. By 1900, in other words, a situation had been created in which the elimination of cheap labor would have involved a readjustment of land values and the entire capital structure of California agriculture. The resulting maladjustment of land values boded ill for the working farmers as well as the farm workers, for the labor of the former was necessarily in competition with the latter.

In the period from 1900 to 1930 a still greater change took place in the character of California agriculture. As the acreage under irrigation trebled, the growth in specialty crops became extraordinary. The vegetable acreage expanded from 32,479 acres to 355,231 acres; from an annual value of $3,000,000 to $60,-000,000. Investment in irrigation systems rose from $19,181,610 in 1900 to $450,967,979 in 1930: an increase of more than twenty-three fold. This expansion came about with scarcely any increase in the number of acres under cultivation; it represented, in other words, an ever-increasing capital investment and a constant intensi-fication of production. During this period, the demand for farm labor increased 133 per cent.

When this remarkable expansion began in 1900, there were proportionately more "white" workers in California agriculture than there had been at any prior period. These workers launched a direct frontal attack on the system of farm labor employment which had developed during the prior thirty years. The first or-ganizational drives, for example, date from 1900. For a brief period it even appeared that California farm employers were pre-pared, at long last, to make some important concessions. But, at this crucial juncture, Japanese immigration began to assume major proportions and the long-deferred readjustment was postponed. The increase in the number of Japanese between 1900 and 1907 not only checkmated the effort to stabilize farm labor conditions and readjust land values, but it brought into being a virulent cam-paign to exclude the Japanese. As in the earlier period, this agita-tion led to the cessation of Japanese immigration and, finally, to its exclusion.

The period from 1908 (when Japanese immigration was sus-pended) to 1917 roughly approximated the period from 1884 to 1900 when large numbers of "white" workers again sought em-ployment in agriculture. But, with the suspension of Japanese immigration, came an influx of several thousand Hindus and the beginning of large-scale immigration from Mexico. Once again, therefore, the opportunity to make an adjustment was lost. The employers failed to act and, on the contrary, began to flood the farm labor market with a wide variety of new recruits, thereby touching off a second organizational drive. This renewed effort to organize farm labor reached its climax with the Wheatland Riot of August 3, 1913, which has been characterized by the La Follette Committee as "a landmark in the history of labor relations in California agriculture." For perhaps the first time, the citizens of California, and of the nation, were made aware of the cancer which had long been festering in the rural areas of the state. The riot, however, initiated a period of violent repression and set back, by a decade or more, the struggle to organize farm workers.

Between 1920 and 1930 the number of Mexicans in the state increased fourfold and, by the end of the decade, some 75,000 Mexicans were taking part in the great migratory farm labor circle. The importation of Mexicans, of course, gave rise to an agitation to place Mexican immigration on a quota basis. Fearful that some such legislation might be adopted, the growers began to import Filipino workers and, by 1930, some 30,000 Filipinos had joined the great procession as it moved from valley to valley, crop to crop. During this decade, also, the first "auto" migrants appeared, traveling about in their jalopies; and families rather than single men began to constitute the mainstay of the migratory movement. Organizational efforts virtually ceased since the farm labor market was flooded to a degree that made any control impossible. The period from 1920 to 1930 was also marked, as the La Follette Committee reported, "by an almost complete cessation of effort by the general public through the state government to ameliorate working and living conditions through protective legislation. It was, in fact, a ten-year lull before a storm which has not yet subsided."

The 1929 depression then ushered in still another chapter. With the onset of the depression, thousands of Mexicans were repatriated by the relief agencies, others huddled in the cities where they had acquired residence and refused to make the annual crop junket, and still additional thousands left voluntarily for Mexico. At the same time, the first of the "dust bowl" migrants began to make their appearance. Between 1935 and 1938 approximately 350,000 Okies and Arkies entered the state and the farm labor market became more disorganized than at any prior period. Relief loads mounted; thousands of workers were stranded in the San Joaquin Valley at the end of the season; and the federal government was forced to intervene. The great farm labor strikes of the decade from 1930 to 1940 reached a magnitude and intensity of violence never previously known. This upsurge of organizational activity was, of course, suppressed with characteristic ruthlessness and brutality. Following the familiar pattern, the influx of Okies and Arkies touched off an agitation against them which, in every respect, closely resembled the agitation against their Oriental predecessors in the migratory movement.

As World War II approached, the Okies and Arkies began to drift into the shipyards and defense plants and, with farm production soaring, the familiar scramble for workers was repeated. First 35,000 Mexican *braceros* were imported under an agreement between the two countries; then prisoners from San Quentin and Folsom Prisons were drafted for service, along with Italian and German prisoners of war, Negro migrants from the South, school children, and the latest arrivals from Oklahoma and Arkansas. In February, 1948, a new agreement was negotiated between the State Department and the Mexican government which permitted growers to recruit Mexican nationals with most of the wartime controls and safeguards removed. Since then, of course, California has been flooded with illegal entrants or "wetbacks" from Mexico who seem to experience little difficulty in finding employment. Currently the growers are experimenting with the use of Navajo Indians and are seeking to negotiate an agreement under the terms of which thousands of workers would be imported from Puerto Rico. Another current project would involve the use of some 14,000 DP's from Europe. When it comes to conjuring up new sources of cheap labor, the California growers are unexcelled; they are probably the world's most resourceful labor recruiters.

The repetitive pattern to be noted in this brief summary is the telltale proof that the farm labor problem in California has become encysted, i.e. embedded, in the very structure of the state's agricultural economy. It will be noted that there is no progression in this record but merely the repetition, at fairly evenly spaced intervals, of an old and extremely depressing pattern. The actors keep changing but the plot is always the same. The warning signs and danger flags have been apparent for nearly a century, yet nothing basic has been done about this problem; in fact nothing can be done about it without a readjustment of the agricultural economy of the state.

THE LABOR POOL

Long before the La Follette Committee made its famous study of farm labor in California, the salient characteristics of the state's farm system had been isolated. In a pamphlet entitled *The Tramp,* Jack London had pointed out why it was necessary to have a "second army"—an army of the chronically unemployed—in California. At an even earlier date, Morrison I. Swift, in a pamphlet on *What a Tramp Learns in California,* published in 1896, gave an illuminating account of California's farm labor system which is as pertinent today as the year it was published. From the reports of the La Follette Committee, however, it is possible to piece together a more systematic and better integrated account of this most peculiar system of farm labor.

The conditions under which California's army of farm workers are employed are quite similar to those which prevail in mass-production industries. Laborers work in gangs with careful supervision by foremen and labor contractors. Workers have a casual relationship to various employers, there being no fixed relationship between the land and the people who work it. Above all, the system is characterized by temporary hiring from a mobile pool of unemployed workers. From the employer's point of view, it is essential that this pool should be large, that the workers should be unorganized, and that the pool itself should be made up of otherwise unemployed workers since only the unemployed will migrate in the manner required by the system. It is also important that the supply of labor should be kept fluid and mobile. To achieve these requirements, the employers must maintain complete control of labor relations. The control must be such, in fact, that hourly and piece rates for various types of employment can be maintained at a uniform level throughout a particular season. For if employers started bidding against each other, on either an area or crop basis, workers could not be routed in the manner desired and wages would tend to rise. Since many crops have the same maturity dates, employers must draw from the same pool of labor and the competition for labor might become ruinous if uniform wage rates did not prevail. High wages, in many California crops, have a tendency to reduce the supply of labor, since the work is so thoroughly undesirable that workers will pick for a short time and then quit. It is important, therefore, that wages be kept at the lowest possible level, not merely to minimize labor costs but to keep workers on the job. Hence from the earliest date associations of farm employers have fixed maximum rates for a wide variety of farm operations without any semblance of collective bargaining. Moreover, they have enforced these rates in the most systematic and thoroughgoing fashion even to the point of invoking governmental controls to hold recalcitrant employers in line. The lack of organization among the workers has not only deprived them of effective representation at these pre-harvest wage-fixing sessions, but it has prevented them from exerting political pressure which might induce government to intervene on their behalf.

This system of employment, which nowadays involves as many as 300,000 workers at the height of the season, is predicated upon a novel relationship of people, both employers and employees, to the land. The typical farm employer in California is not, in any sense, a "farmer." One of the leading California farm journals stated this new relationship with typical bluntness: "The incidents of husbandry, the family-sized farm with all its pastoral glamor, is a lovely idyll—elsewhere than most sections of California." The large shipper-growers "farm by phone" from headquarters in San Francisco or Los Angeles. Many of them travel, nowadays, exclusively by plane in visiting their various "operations." Although the relationship between these employers and their industrial colleagues is most intimate, their relationship to the land is as casual as that of the migratory workers they employ. "The term 'farm,'" as one expert has said, "is long since obsolete in respect to the highly specialized agricultural occupations of growing such products as citrus fruits, walnuts, avocados, and a large number of other specialty crops grown on the Pacific Coast." It should be noted, moreover, that over half the farms of California employ no farm labor whatever. The farm labor problem, therefore, is intimately related to the scale of farm operations.

Not only does the demand for farm labor vary from month to month—from 50,000 in January to 300,000 in September—but the demand constantly shifts from crop to crop, from one area to another. Crops which require the most labor frequently have the same maturity dates. Many of these crops must be harvested at a particular time, not only because they are perishable but for other reasons as well. For example, grapes picked before September 1st can usually be converted into raisins in a period of two weeks; but, for reasons not yet known, grapes picked after September 15th usually take a month to be converted into raisins. As the farm labor system operates today it assumes, in fact it is based upon, underemployment and unemployment. Even if the employers made a serious effort to rationalize labor operations, there would still be both underemployment and unemployment. As the long and painful record shows, the labor market has always been characterized by purposeful disorganization and, for most periods, by a chronic oversupply of labor.

Although there are many features about this peculiar system of farm employment that the employers themselves do not like, it has three distinct advantages from their point of view. Since the nature of the system makes the organization of farm workers a most difficult undertaking, the employers are invested with exclusive control over the incidents of employment, including wages and working conditions. By keeping the labor market unorganized, labor costs can be minimized. Also, under this system, the employers have been able to minimize the cost of transporting workers, providing decent housing, and planning and coordinating labor operations which they might be compelled to undertake if the labor market were organized. Under this system, in short, "labor is an unemployed pool available on call, much in the manner of water or electricity." It would serve no purpose here to describe the miserable working conditions which are part and parcel of this system of employment and which have prevailed, with little mitigation, for over seventy years. A survey of the unnumbered reports and investigations which have been made over a long period of years convinced the La Follette Committee that the record "discloses, with monotonous regularity, a shocking degree of human misery."

Just as this system of farm employment has become part of the basic structure of California's agricultural economy, so it has produced a set of attitudes and assumptions on the part of employers which have long since become part of the established folklore of the state. There was a time, when the crudities of this system first became apparent, when farm employers were conscious of its abnormalities, i.e. they were then quite willing to admit that the system was socially inefficient and that it worked many injustices. At that time they defended the system by saying that only Orientals could be obtained as farm workers and that, later perhaps, when more "white" workers were available, they would undertake to improve working conditions. The point to be noted is that there was once a time when farm employers were not wholly insensitive to the shortcomings of this system; but, by the middle 1920's, the concept of abnormality had almost completely disappeared. By then, as one observer has pointed out, "California agriculture was declared *by nature* to be such as to demand a permanent supply of itinerant laborers." Once the dogma was established that only Mexicans, Filipinos, and other minorities could perform the "stoop labor" jobs, it became possible to dismiss suggestions for improvement as "idealistic," "utopian," and "impractical." This dogmatism has been encouraged by the fact that no representative body of public opinion has ever seriously challenged the underlying assumptions; and also because the farm labor pool has never been in a position "to talk back" to the employers or to state its side of the case. Farm employers in California have become so "used to" this system that they regard all its incidents as part of the natural order of the universe, and with this acceptance has come a firm belief in all the clichés, shibboleths, and stereotypes which have grown up with the system itself. It would be difficult, therefore, to find in the whole range of American employer groups a more thoroughly opinionated and arrogant lot than the men who boss the California farm labor system. They not only know all the answers but dismiss as subversive any questions that are asked.

Underlying the attitude of the California farm employers, however, is a sense of fear and guilt. They are not unaware of the inconvenience and misery, the hardship and suffering, which is implicit in the system itself. But they sense, even if they will not willingly admit, that a readjustment of this system would involve a readjustment of the entire agricultural economy. Hence they are driven to defend the system and its consequences much as slave-owners were driven to defend chattel slavery. Suggestions for the improvement of labor camps are brushed aside with stories about farm workers who urinate in kitchen sinks, cut holes in the floor for toilets, and chop up the partitions for kindling wood.[1] I had occasion to deal with these growers for four years as Commissioner of Immigration and Housing in California in charge of the inspection of some 5,000 agricultural labor camps, and I can testify from bitter personal experience that for every suggestion they have a time-honored rationalization; for every criticism a hoary and preposterous fable.

The sense of fear and guilt also arises from an acute realization that there are many phases of California agriculture that will not bear public scrutiny, such as certain phony marketing agreements, the lush subsidies, the wanton destruction (at various periods) of tons of food, the rigging of market prices and so forth. Even the typical farm labor employer in California is aware of the fact that there is something grossly inconsistent about a dispensation which permits government to pay fancy parity payments to growers, but holds that it is heresy to suggest that government might provide decent camps for farm workers. The sense of fear is implicit in the vulnerability of California agriculture to labor organization. Although the organization of farm workers presents many difficulties, if they were once organized they would have a tremendous power; and it is the realization of this fact that haunts the employers. Many California growers live in mortal fear of any inter-

[1] See, for example, the comments of one grower as reported by Stanton Delaplane in the San Francisco *Chronicle*, March 26, 1948.

ruption in the careful schedule of labor operations upon which they may have gambled a fortune.

On the other hand, farm workers by reason of the nature of their employment are prone to form an extremely unflattering picture of their employers. Without status, security, or protection against the hazards of employment, they have nothing whatever to look forward to in the way of advancement. They know all about the hardships of their employment but they know nothing about the problems of their employer since there is no occasion whatever for them to become familiar with these problems. If the employer lives in fear of strikes, a break in the market, or a stretch of bad weather, the farm laborers fear that next week they will be broke and without a roof over their heads. The employers' concern about time is matched by their indifference. He wants the crop harvested as rapidly as possible; they would like to see the job last another week. He does not know them and possibly will never see them again; they do not know him and probably would not care to. If he is inclined to think that they are an inefficient, happy-go-lucky, worthless lot, without ambition and indifferent to their fate, they are prone to believe that he is a hard-boiled bastard against whom the use of any tactic is justified. If ever a system was calculated to make for bad labor relations, it is this system of farm labor employment in California.

THE SYSTEM OF CONTROL

As the migratory labor pool expanded—from 119,800 in 1920, to 190,000 in 1930, to nearly 350,000 in 1939—it became necessary for California farm employers to develop a network of organizations by which an unchallenged control of the labor market could be maintained. There are "farm" organizations as such in California—the Grange, the Farm Bureau, and the Farmers' Union; but the organizations to which I refer are those primarily concerned with labor problems. Today these various farm employer associations represent a formidable structure of power. At the base of the structure are the various area and commodity organizations whose control over labor policies is such that they are able to force every producer of a particular crop, or all producers in a certain area, to adhere to their policies. Ostensibly these associations "coordinate" policies, but actually they formulate and enforce labor policies. The farmer employer does not decide what wages he can pay; he is told what wages he must pay. The association recruits labor, supervises its distribution, and determines the conditions of employment.

But, since all farm employers tap a common pool of labor, it also became necessary to coordinate the policies of the local associations on a state-wide basis. This coordination has been effected since 1920 through the powerful Agricultural Committee of the State Chamber of Commerce. Through this committee, also, agricultural labor policies are coordinated with those of industry. In the 1930's it became necessary to have still another organization, one primarily designed to suppress strikes. This organization, the Associated Farmers of California, came into being on March 28, 1934. Up to this time, the organization of farm workers had never reached a point of sufficient effectiveness to require continuous attention; it was the great farm strikes of the thirties that brought the Associated Farmers into being. Devoted to "vigorous and curative action," the Associated Farmers represents the police power of California agriculture—its enforcement arm.

With the formation of the Associated Farmers, one could say that California agriculture had achieved total integration. As a structure of power, this integration commands admiration. It is no small feat to have been able to organize farm employers, producing so many different crops, with so many special labor problems, in this thorough-going and inclusive manner. Carefully coordinated from the top, the organizational network functions with clocklike precision, policing entire crop industries, enforcing uniform decisions, holding recalcitrant employers in line. When the need arises, the full weight of this powerful apparatus can be brought to bear upon any threatened sector of the agricultural front with crushing force and effectiveness. Local, county, state, and, on occasion, federal officials jump when the Associated Farmers crack the whip.

There is no parallel in any state for this interlocking network of farm employer organizations which represents a most unique combination of social, economic, and political power. Every weapon in the arsenal of anti-unionism has been used by these employer organizations at one time or another. For example, the Associated Farmers has used undercover agents and *agents provocateurs;* it has blacklisted employees; devised variants on the Mohawk Valley formula; broken countless strikes; stimulated "direct action" campaigns; maintained a state-wide intelligence service; and has frequently subverted the machinery of law and order. Elaborate supporting documentation can be found for each of these statements in the findings and report of the La Follette Committee. With the formation of the Associated Farmers, farm labor became "totally engaged" with farm employers in California.

The existence of this network of employer organizations has had important social and political consequences. It has meant, for example, that employer strength has not been broken up into farm and non-farm categories but has been consolidated into a single apparatus. Social power in California is not divided between labor, industry, and agriculture; it is divided between labor and industry. In other states the farm groups often mediate between industry and labor or both sides will seek to win the farm support; but in California industry and agriculture are one.

There are, of course, thousands of bona fide farmers in California and they are not without power and influence; but, by an overwhelming margin, power nowadays resides in the large-scale industrialized farms. Farm employer groups in California will automatically support urban industrial groups on any major labor issue. Any effort to reform the farm labor system, therefore, must cope with this formidable structure of power. The farm labor system is not, as I have indicated, without its inherent weaknesses; but these weaknesses are more than offset by the complete organization of farm employers and the total absence of effective farm labor organization.

In California, moreover, there are various industries which represent a fusion of industry and agriculture or which belong in both camps. The canning, processing, drying, sugar beet refining, and cotton-ginning industries belong in this category. The labor problems of these industries are closely related to those of agriculture proper. But, since the peak periods of employment in these industries correlate with those of agriculture, there is little overlapping in employment. Hence the dichotomy which has so long existed between "field" and "shed" workers. These intermediate industries, of course, have their own employer organizations which are often controlled by the same interests that dominate the various farm employer organizations.

Vast industries have developed in California to process, distribute, and transport farm products; to provide electrical power, supplies, machinery, boxes, ice, and paper. The fruit and vegetable canning and drying industry alone has 70,000 employers and produces products of an annual value in excess of $250,000,000. One-third of all outbound shipments from the Port of San Francisco in 1937 consisted of agricultural products so that shipping, too, is intimately related to agriculture. In the same year, agricultural products made up nearly 30 per cent of the freight revenues of the four major rail lines operating out of the state. In 1929, California agriculture consumed almost one-third of the total electric light and power used on American farms. In 1937, the canning industry of the state spent nearly $50,000,000 for cans, approximately $20,000,000 for sugar, $5,000,000 for paper labels, and $8,000,-000 for fiberboard cases. A large part of the outlay of the lettuce industry is for lumber, paper, labels, ice, gas and oil, nails, and transportation costs. Salinas, the lettuce capital of the state, with a population of 21,000, manufactures more artificial ice than San Francisco. A host of industries, therefore, have an enormous stake in California agriculture. If cotton is left unpicked, if fruit rots on the trees, if vegetables are not harvested, these related industries are immediately affected.

Needless to say the industries directly dependent upon agriculture support the labor policies of the farm employer organizations. It is not only important to these industries that crops should be harvested but they have a vital stake in seeing to it that the charges for goods and services which they levy on agriculture are kept at a relatively inflexible level during periods of declining prices. It is also to their interest to keep farm wages as flexible as possible so that, in periods of declining prices, the adjustments will be at the expense of labor. The less pressure from labor on farmers for more wages, the less pressure from farmers on these industries to reduce their charges for goods and services. This same relationship exists elsewhere in American agriculture, but what makes California unique, in this respect, is the diversity of the industries which occupy a more or less parasitic relationship to agriculture. Only a limited number of satellite businesses can be developed around the production of corn, wheat, and cotton; but there is almost no limit to the number of subsidiary businesses which can be developed out of California's diversified agricultural production.

THE PLOT NEVER CHANGES

On October 1, 1947, 1,500 farm workers, all employees of Joe Di Giorgio's great 20,000-acre farm factory in Kern County went on strike: the issue was union recognition. Within a matter of weeks, 20 strikers had been arrested and, to secure their release, the union had been required to post more than $70,000 in bail bonds. Then the usual evictions began as the families of the strikers were ordered out of company houses. A fake "Citizens' Committee" was appointed at the suggestion of the Associated Farmers which, of course, issued a report exonerating the employer of all responsibility for the strike. Despite the fact that the strikers were excluded from the provisions of the Taft-Hartley Act, being "agricultural employees," an injunction was nevertheless secured which prohibited all secondary picketing. The California Committee on Un-American Activities, following a hurried investigation, suggested that "Communist" influences might be behind the strike although the union, an affiliate of the American Federation of Labor, bars Communists from membership. Some 300 Mexican nationals, recruited as strikebreakers, were rounded up by the immigration authorities but the company promptly brought in several truckloads of scabs from Texas. Then, on May 18, 1948, the chairman of the strike committee was shot in the head when a volley of bullets riddled a house in which the committee was meeting.

Now just what is the nature of this "farm" where the strike occurred? The farm is one of several properties owned by the Earl Fruit Company which is in turn owned by the Di Giorgio Fruit Corporation which is owned by Joe Di Giorgio. In fact, Earl Fruit Company operates 27 farm properties in California and leases 11 additional properties. It also purchases a large amount of fresh fruit every year from other growers. It owns 11 packing houses and packs and markets about 1,000 cars of fresh fruit annually. It also owns a 95 per cent interest in the Klamath Lumber and Box Company which can turn out 25,000,000 feet of lumber a year for boxes and crates. It also owns two wineries, one of which is the largest in the nation, and it is building a third winery which will have a storage capacity of 10,000,000 gallons.

But Joe Di Giorgio is a fruit merchant as well as a fruit grower. So the Earl Fruit Company owns the Baltimore Fruit Exchange and has important holdings in fruit auction houses in Chicago, New York, Cincinnati, and Pittsburgh. In 1940, the company employed 2,887 farm workers and had an annual payroll of $2,400,000. To provide accommodations for its regular employees, some 350 "cottages" are maintained with bunkhouse facilities for 2,000 additional seasonal workers. Through still another subsidiary, Di Giorgio owns 13,833 acres of farm land in other states. In 1948, the Di Giorgio Fruit Corporation reported sales and commissions cf $11,837,545.55, and net earnings for the year, after taxes, cf $247,701. Yet by masquerading as a "farm," this huge operation enjoys complete immunity from virtually every form of state and federal social legislation.

The "home farm," where the strike was called, has an interesting history. This property first came to the attention of Joe Di Giorgio, a Sicilian immigrant, in 1915 when he discovered, more or less by accident, that figs produced in this particular area do not have to be protected against night and early morning dew. Most of the 20,000-acre tract was purchased for about $90 an acre (it is now worth $2,500 an acre); and was improved at a cost of $275 an acre (the improvements would now cost $850 an acre). From this base of operations, Di Giorgio's interests have expanded to the point where he now ships around 10,000,000 boxes of fruit and 500,000 packages of vegetables annually and employs, in all his operations, around 10,000 employees. In 1946 the parent company and its affiliates paid the federal government $1,614,817 in taxes, yet the home property itself is still regarded as a "farm" and is therefore placed in the same legal category as a 40-acre orchard owned and operated by a farm-family in the immediate vicinity.

On the scientific and technological side, the Di Giorgio operation is a miracle of efficiency and large-scale organization; but it cannot be given a high rank in terms of social efficiency. In the first year of the Di Giorgio strike, Stanton Delaplane of the San Francisco *Chronicle*, in a series of excellent articles, reported that, within a few miles of Fresno, there were children of farm workers who went without food for twenty-four hours at a stretch. In the Coalinga-Huron district, 1,700 *families* were destitute in the winter of 1948, existing on handouts from the Red Cross, the Salvation Army, and the local service clubs. "The first thing you notice at the Firebaugh school," he reported, "are the shoes. Eight-year-olds walking around in sport wedgies. Youngsters with their father's work shoes. High shoes, low shoes, button shoes; but not children's shoes." Headlines from the valley newspapers tell the story: "Farm Work Sought for 5,000 Without Jobs"; [2] "Needy Migrants Create Relief Problem"; [3] "Polio, Diarrhea Cases Increase in Kern County"; [4] "Children Living in Labor Camps Are Called 'Lost' "; [5] "Relief Agencies Foresee New Wholesale Want in West Side Labor Camps"; [6] "Migrant Influx Jams Hospital in Bakersfield"; [7] "Kern Crime Is Blamed on Farm Worker Poverty"; [8] "Valley Officials Will Tackle Farm Hunger Problem"; [9] "Labor Camp Rent Shows Increase of 50 Per Cent"; [10] "Labor Camps in Valley Are Found in Poor Condition." [11]

[2] Fresno *Bee*, January 9, 1948.
[3] *Ibid.*, February 11, 1948.
[4] *Ibid.*, November 16, 1948.
[5] *Ibid.*, December 11, 1948.
[6] *Ibid.*, December 8, 1948.
[7] *Ibid.*, December 16, 1948.
[8] *Ibid.*, February 17, 1949.
[9] *Ibid.*, March 10, 1949.
[10] *Ibid.*, March 19, 1949.
[11] *Ibid.*, March 17, 1949.

These headlines read like the latest "news"—as though they were heralding some novel and emergency situation; yet, in the Fresno *Bee's* "Fifty Years Ago" column of August 29, 1948, one could read items which indicated that, on August 29, 1898, exactly the same conditions had prevailed. They have prevailed, in fact, for nearly a hundred years. Farm crops produce an annual income for Fresno County of $180,000,000. With this in mind, it is, indeed, disturbing to read an editorial in the Fresno *Bee*—the leading newspaper of the San Joaquin Valley—in which the community is congratulated because the number of unemployed migrant workers has declined from 22,000 to 18,000 and then to read this statement: "So seasonal unemployment is something to be expected. It definitely is nothing to cause unusual concern hereabouts." [12] In other words, it has always existed; therefore, it is "normal," something to be expected and accepted, with never a hint or a suggestion that anything might be done, if not to "solve" the farm labor problem, then to mitigate the hardship and suffering which it causes. In point of fact, however, the solution to this problem, in large part, has been known for many years. If the measures which the La Follette Committee recommended to Congress on October 12, 1942, were enacted, the worst aspects of California's farm labor problem could be removed. But these recommendations, based upon one of the finest and most thorough-going of all congressional investigations, are still gathering dust in the archives of Congress and the Californians are still discovering, every winter, that they have a migrant farm labor problem.

OF THE VALLEYS

It is the fusion of agriculture and industry that so largely accounts for certain intangible qualities that one senses in California. The relative absence of "rural" opinion, for one thing, creates curious social effects. No matter how familiar one may be with "rural" California, it is always rather surprising to note the manners and appearance of the gentry who step forward to speak in the name of "the farmers" at legislative hearings in Sacramento. These men

[12] *Ibid.,* February 15, 1949.

are "operators," not farmers. The existence, also, of a flourishing "rural" underworld of lotteries, gambling, cockfights, and prostitution creates a most baffling impression. It is as though phases of city life had been grafted on rural stock. During my term as Commissioner of Immigration and Housing in California, I had occasion to see portable houses of prostitution making the rounds of the labor camps in the Delta district: trailers, with four and five girls, hitched on to cars and drawn to first one camp and then another. I have watched fabulous crap games and cockfights in labor camps in which 700 to a thousand workers were living. I have flushed scores of migrant families out of abandoned barns, from beneath bridges, and from the zaniest canal-bank habitations that one could possibly imagine. On tips from anonymous sources, I have gone in search of labor camps in areas where there were no visible signs of tents, cabins or shacks and, eventually, have turned up as many as 200 and 300 families in artfully concealed camps. One can travel the length of the San Joaquin Valley, at the height of the season, on the main highway without being aware of the fact that tens of thousands of migrant workers, an army of 200,-000, are somewhere camped, somewhere at work. But, once you know that this curious "hidden" world exists, you are forever conscious of it and your eyes seek out the evidence that this phantom army is there, in the vineyards and orchards, in the camps and shacktowns.

At the other extreme, one comes to know that interesting figure, the California "farm industrialist" who wears a neat Stetson, travels in an airplane, and has the breezy manners and the swagger of a Texas "cattle king." These are the men who have luxurious homes in San Francisco and Berkeley as well as gracious mansions in the valley towns; men whose lives are spent in motion between farm and city, and who are not really of the farm nor of the city. California novels are full of these "mixed" characters who are so difficult, at first, to place, so hard to identify, for they have no precise counterpart in American life, with the possible exception of the rich Texas cattle and oil barons. These men are somehow as conspicuous in the country as they are in the city. If they affect in the city some eccentricity of manner or appearance which identifies them as being not quite of the city, they also affect an ostentatiousness in the valley towns which sets them apart from the working farmers and the townspeople.

To be in the major valley towns at the height of the season is to become acutely aware of the climax of a process of production which is neither agricultural nor industrial. The canneries work shifts of girls and women around the clock; schools of trucks zoom along the highways, day and night; long lines of refrigerator cars are shunted about in the railroad yards as they are iced and loaded; and the taxi dancehalls in the skidrow sections, charging ten cents for a dance that lasts precisely one minute, do a fabulous business. The nights do not seem to be nights at all, for they teem with life and activity and commotion. On the outskirts of the towns are the seasonal junkyards in which migrant families trade and barter for old cars, pots and pans, brass bedsteads, and old clothes. Each of these towns has at least one luxury hotel at whose air-cooled bar the Stetson-hatted gentry foregather. In an upstairs suite, one can usually witness a poker game with stakes running into the thousands of dollars. Everyone gambles; even the poor play the slot machines, the grab machines, the consoles, and the punchboards, as they pour a torrent of nickels, dimes, and quarters into the ever-present juke-boxes. Gambling, implicit in this form of "farming," is the order of the day. Coming to a frenzied climax in September, this curious "rural" life of the valleys has a flavor all its own, a flavor that is strictly Californian.

10

PERILOUS REMEDIES FOR
PRESENT EVILS

◇◇

SINCE CALIFORNIA was admitted to the union the nation has been vaguely aware of something "different" and "peculiar" about the politics of the state. In the 1870's the nation, disturbed by the rise of Kearneyism in California, seriously debated whether California had "gone Communist." Then for nearly seventy years California's phobia about Oriental immigration continued to upset and confuse the nation. On the first Tuesday of November, 1916, the American people went to bed thinking that Charles Evans Hughes had been elected president only to discover, on Wednesday morning, that California had re-elected Woodrow Wilson. In the 1930's Kearneyism found its almost exact counterpart in Mr. Upton Sinclair's amazing campaign to end poverty in California. These and other episodes have served to create in the popular mind, as Harold F. Gosnell has said, an impression that "California has long been a state apart from the rest of the union, where the exotic, the unusual, and the peculiar grow in profusion —both in nature and in human institutions." What is there, then, about the Californians that has made them so impatient, as Lord Bryce put it, "with the slow approach of the millennium" and so consistently eager "to try instant, even if perilous, remedies for present evils"? In this chapter I want to focus attention upon certain key episodes in the political history of the state for the purpose of illustrating the peculiar dynamics of California politics.

KEARNEYISM IN CALIFORNIA

In the late 1870's California was seething with social and political discontent. As the period of "free mining" came to a close, the yield of gold decreased, wages fell, land values soared, and an ever-increasing proportion of the state's income went to those who charged rent, speculated in land values, and collected interest and royalties. The completion of the Central Pacific, as Henry George had foreseen, had not made for universal prosperity; on the contrary, it had increased the power of the wealthy classes and undermined the position of the poor. Resentment against the railroad ran high in both urban and rural areas; no state, as Lord Bryce said, had ever been so completely at the mercy of a single corporation. Mining stocks declined in value, job-seekers poured in from the eastern states to demoralize the labor market, and, in 1876, the peak of Chinese immigration was reached with an influx of 22,493 Chinese. The unemployed in San Francisco were mostly single men, ex-miners and former railroad workers, who cherished a deep resentment over the disappearance of the "flush times" of the gold rush decades. The farmers of the state, hard hit by drought and deflation, the exactions of land speculators and excessive freight rates, were also in a rebellious frame of mind. In other words, a variety of factors had combined to produce a situation which, in the words of Lord Bryce, was both "peculiar" and "dangerous."

These conditions were certainly not unique in the America of 1877, although there was probably more reason for discontent in California than elsewhere; but the uniqueness of the California situation consisted in the absence of settled and well-established forms of social organization through which this pervasive discontent might have found a more "normal" expression. The evolution of social forms in California had not kept pace with the growth in population. California's newly made millionaires, for example, lacked a fully matured sense of social power. Grown rich through lucky speculations and "strikes," many of these men displayed their wealth with a vulgar, provocative, and unbecoming ostentation; as a group they were lacking in any sense of civic leadership and social responsibility. "There was, therefore," wrote Lord Bryce, "nothing to break the wave of suspicious dislike" which the unemployed felt toward the rich. The millionaires were quite close, it should be remembered, to the unemployed in temperament, social origin, and former circumstance. A lucky turn of the wheel of fortune had made some rich and others poor, and the memory of their common origin rankled among the poor. Not only was there an absence of internal stabilizing factors, but the remoteness of the state removed it from the steadying influence of older sections. "San Francisco," as Lord Bryce said, "was a New York which has got no New England on one side of it, and no shrewd and orderly rural population on the other, to keep it in order."

There was, however, a special factor underlying the rise of Kearneyism in California. As a boy of nineteen, Henry George had boarded a schooner in San Francisco to take part in the Fraser River gold rush. One day, on the deck of the schooner, he had asked an old miner what harm the much-maligned Chinese were doing in the mines if, as it was said, they were permitted to work only the "cheap diggings." "No harm now," replied the miner; "but it will not be always that wages are as high as they are today in California. As the country grows, as people come in, wages will go down and some day or other white men will be glad to get these diggings that the Chinese are now working." By 1877 this "some day" had arrived. Sharply set apart by racial and cultural differences, the Chinese were a natural target for rising social discontent.

In July, 1877, a meeting was called in San Francisco to express public sympathy for the railroad workers who were on strike in Pittsburgh. Frightened by some strong language used at this meeting, and uneasy about the situation, the nabobs of the city formed a Committee of Public Safety patterned after the famous Vigilante Committee of 1856. Raising a war chest of $100,000, rifles and ammunition were purchased for a "pick handle brigade" that patrolled the streets. Among these hastily recruited patrolmen was one Dennis Kearney, an Irish immigrant, formerly a seaman, then a drayman, who had recently graduated from San Francisco's Lyceum of Self-Culture where he had studied "elocution." Far from allaying popular discontent the business community's show of force produced exactly the opposite reaction. It was "this uprising among the society-savers," as Henry George, Jr., later observed, "that tended to bring to a head discontent among the disorganized working classes."

An earlier Workingmen's Party, an off-shoot of a national movement led by Wendell Phillips and Peter Cooper, had called a state convention in 1867 and, for some years, had taken an active part in state and local politics. Kearney, incidentally, had been denied admission to this party. Re-formed as the People's Independent Party, the workingmen had elected their nominee to the Supreme Court in 1874 and had been influential in the election of William Irwin, the Democratic nominee for governor, who had promised to call a state constitutional convention. On September 12, 1877, the leaders of this earlier movement had come together, adopted a platform, and reassumed the name of the Workingmen's Party. Four days later the first of the famous "sandlot" meetings was held in San Francisco. Noting these developments, the nabobs demanded that speakers at the sandlot meetings be arrested, and induced both the city and state authorities to adopt "gag laws" aimed at the suppression of free speech and the right of assembly. So severe was this repression that the first state meeting of the Workingmen's Party was held in secret session. The gag laws, as Miss Eaves noted, "were peculiarly out of harmony with the liberty, often amounting to license, that often characterized the speech of the early Californians."

At a sandlot meeting on October 5th, 1877, the ex-vigilante Dennis Kearney mounted the platform to recommend "a little judicious hanging" for the Nob Hill millionaires and raised the famous slogan "The Chinese Must Go!" Within a few months, Kearney was in undisputed control of the Workingmen's Party from which he had been previously excluded. The platform of the new party, adopted on May 16th, 1878, called, among other things, for the adoption of the eight-hour day; direct election of United States Senators; compulsory education; abolition of contract labor on public works; state regulation of banks, industry, and railroads; and a more equitable tax system.

The new party drew most of its strength from the Democratic Party, a circumstance which naturally pleased the Republicans. On the other hand, the Democrats offered only a token resistance since they ultimately hoped to capture control of the new party. But once the call to the state constitutional convention was issued, the "sound" elements in both major parties experienced a swift reversal of attitude. To prevent the Workingmen's Party from capturing control of the convention, these elements prepared a slate of "non-partisan" delegates to consolidate their strength. Of the delegates finally selected, 51 belonged to the Workingmen's Party (one of these delegates had been a member of the Paris Commune); 78 were "non-partisans," 11 Republicans, 10 Democrats, and 2 Independents. The Workingmen's delegates had been elected, incidentally, at an expense of precisely $300. Although not a majority, the Workingmen's Party exerted a great influence in the convention and many of its proposals were adopted. Later, however, the radical provisions of the 1879 constitution were robbed of meaning by a familiar process of judicial nullification and legislative sabotage.

Among Kearney's cohorts in the plot to capture control of the Workingmen's Party was Charles de Young, publisher of the San Francisco *Chronicle*, a man with an extraordinary talent for gossip and slander. After the constitutional convention, De Young broke with Kearney, formed a new party of his own, and devoted his vitriolic talents to the destruction of Kearney and the Rev. Isaac Smith Kalloch, candidate of the Workingmen's Party for mayor of San Francisco. Digging up some unsavory episodes in the career of the Rev. Kalloch, De Young devoted columns of the *Chronicle* to the systematic defamation of Kalloch, who was then the best known minister in San Francisco. Kalloch in turn proceeded to denounce the "bawdy house breeding" and "gutter-snipe training" of Charles and "Mike" de Young. Charles de Young then attempted to assassinate Kalloch, and nearly succeeded, only to be murdered by Kalloch's son. In the wake of this political split and its sensational denouement, the Workingmen's Party rapidly disintegrated. However, in the election of 1880 the candidate of the Workingmen's Party and the candidate of the New Constitution Party (de Young's party) polled a combined vote nearly twice that of the Republican candidate for governor but the split enabled the latter to win.

Kearneyism provides a striking illustration of certain persistent trends in California politics. Note the quickness with which this movement of social protest precipitated a major political upheaval. In the space of less than three years, the new party had forced the adoption of a new constitution and had upset the balance of power within the state. It was the "surprising ease and swiftness" of this movement's success that most impressed Lord Bryce, as he was also impressed by the fact that the movement "fell as quickly as it rose." The Workingmen's Party elected 10 senators and 16 assemblymen in 1880; 10 sensators and 4 assemblymen in 1881; and "shortly thereafter passed out of existence." By 1883 the party was no longer in existence.

Both the ease with which this movement arose and the swiftness with which it collapsed were essentially aspects of the feebly rooted character of social organization in California. There was nothing to break or check the rise of the movement and, by the same token, there was nothing to hold it together once its initial objects had been achieved. Deeply rooted political traditions would have kept such a movement in check in other states; but the traditional factor was almost wholly lacking in California politics. Precisely the same factors account for three subsequent political upheavals in the state. Before considering these upheavals, however, it would be well to note Miss Eaves' summation of Kearneyism. "The Workingmen's Party," she wrote, "owed its success to a *spontaneous* uprising of the wage-workers expressing itself in a way with which they had become familiar," namely, independent direct political action.

THE UNION LABOR PARTY

Shortly after the "mutual cessation of hostilities" brought an end to the bitter waterfront strike of 1901, a municipal election was held in San Francisco. Just as the "gag laws" of 1877 and the show of force by the business community had greatly stimulated the rise of class consciousness which had precipitated the revolt of the Workingmen's Party, so the formation of the Employers' Council and the use of violence in the 1901 strike brought about a new political upheaval. The Union Labor Party, which was born of this strike, was not officially recognized by the trade union leaders at the outset; on the contrary, it represented, in Miss Eaves' phrase, "a *spontaneous* expression of dissatisfaction."

At the time the waterfront strike was called on July 30, 1901, nothing had been said or suggested about the formation of a labor party; but, shortly after the strike was concluded, the Union Labor Party had been formed and its nominee, Eugene Schmitz, had been elected mayor of San Francisco by a landslide vote. In the elections of 1902 the new party nominated and elected 1 state senator, 7 assemblymen, the superintendent of schools in San Francisco, and 2 Congressmen. In 1903 Schmitz was re-elected mayor by an even larger majority than in his first campaign. In this election, also, the Union Labor Party elected 1 judge, 3 assemblymen, and 3 state senators but lost the 2 congressional seats which it had won in the prior election. Then, in 1905, the party swept the city election and captured control of the entire apparatus of city government in San Francisco.

With the sweeping victory of 1905, Schmitz, by profession a bassoon-player, and his principal backer, Abraham Ruef, threw discretion to the winds and began the systematic plunder of San Francisco. The motto of the administration seemed to be, it was said, "Encourage dishonesty, and then let no dishonest dollar escape." Schmitz and Ruef went into partnership with dishonest contractors; sold privileges and permits; extorted money from restaurant and saloon owners; levied assessments on municipal employees; shared in the profits of houses of prostitution; blackmailed gamblers; sold franchises to corporations; leased rooms for municipal offices at exorbitant rates, and then compelled the lessors to cut them in on the profits; and, generally, "took bribes from everybody who wanted an illegal privilege and was willing to pay for it." The earthquake and fire of April 18, 1906, delayed the impending investigation of this notoriously corrupt regime, but the graft prosecutions finally got under way the next year and, eventually, Ruef and Schmitz were indicted.

Following the graft prosecutions, however, a labor leader, P. H. ("Pin-Head") McCarthy was elected mayor in 1910. Thus, for nearly a decade, the Union Labor Party controlled the office of mayor and most of the city offices in San Francisco as well as the San Francisco delegation to the state legislature. The Union Labor Party, like the Workingmen's Party, had come from nowhere; it had risen to sudden power and influence without prior preparation and in the teeth of opposition from the established trade union leadership. In fact, its formation was so spontaneous that no one quite knew how the party had come into existence. Like the Workingmen's Party, the Union Labor Party fell into the hands of demagogues and grafters for there was much circumstantial evidence to indicate that Kearney, the ex-drayman and elocutionist, was as corrupt as Eugene Schmitz, the bassoon-player. Both movements, it will be noted, survived for about the same period of time. For the Union Labor Party really ceased to exist, as a labor party, after the 1905 election. In short, the episode of the Union Labor Party was a re-enactment, strikingly similar in origin and impulse, to the Workingmen's Party upheaval of 1878.

KEARNEYISM RE-ENACTED

Over a period of a great many years, the Democratic Party functioned in California merely as a convenient foil for the Republicans. The Republicans held the governorship from 1898 to 1938; elected a majority of the assembly from 1898 to 1937; and have controlled the state senate from 1898 to the present time. In part, California's Republicanism reflected the fact that so much of the migration to the state had been from traditionally Republican areas; migration from the Southern states was only important during the gold rush. Not more than 5 per cent of the state's population in 1900 had been born in the Southern states by comparison with 20 per cent in 1850. From 1860 to 1932, the Democrats were at a distinct disadvantage since the Democratic registration was never more than 20 per cent of the Republican registration.

During all this time, however, there was an invisible third party in California—the Southern Pacific Company. The Southern Pacific had first entered the political arena in 1879 when a state commission had been authorized to regulate the railroads which, in California, meant the Southern Pacific. By bribing two of the three members of the original commission, the company had made a mockery of state regulation, and through its control of state politics the same mockery persisted after 1879. In all respects the Southern Pacific was essentially a third party. "This third party," as one observer said, "has the usual attributes of a political party, the same apparatus and adherents. It selects these from both parties, but mostly from the party in the majority. Whether they call themselves Republicans or Democrats, and however they divide or contend on party issues, they move as one man in the cause of the railroad against the people."

Until the completion of the Santa Fe line in 1886, the Southern Pacific had held a monopoly on rail transportation to and from, as well as within, the state. By a variety of devices, the company had eliminated any effective competition on the Sacramento and San Joaquin Rivers and, through an alliance with the Pacific Mail Steamship Company, ocean competition was kept at a minimum. There was only one railroad line across the Isthmus of Panama and this line was controlled by Pacific Mail which had given the Southern Pacific the exclusive privilege of through billing. As a matter of fact, total commodity shipments each year, by way of the Isthmus, had declined from $70,202,029 in 1869 to $2,506,177 in 1884. If ever a state, therefore, was at the mercy of one corporation, California was at the mercy of the "octopus"—the Southern Pacific.

Strange as it may appear, however, the "octopus" had many powerful friends and allies in California. No rail system had a greater variety of rates than the Southern Pacific and the variety of rates reflected the range of discrimination which the company practiced. Through rebates, concessions, and special rates, it discriminated against certain customers and at the same time favored other customers in such a way as to build up a network of powerful political alliances. "Where there is a constant demand for favors," writes the historian of the company, "there is likely to be discrimination" and nearly every economic interest in the state sought the favor of the company. In the absence of effective regulation, the company favored the shippers of raw materials from California, for it needed traffic on the long eastern haul, and likewise favored the terminal ports, for it always feared the possibility of ocean competition. It would be difficult, in fact, to find a parallel for the manner in which the company had entrenched itself in California politics.

The revolt against the Southern Pacific began in Southern California where, in 1906, a group of reformers had succeeded in winning a municipal election in Los Angeles. Elated by this victory, they formed the Lincoln-Roosevelt League with the aim of capturing control of the state government. Unlike San Francisco, Los Angeles had never been ruled by a city boss for the basis of bossism did not exist. In San Francisco, on the other hand, the elements of bossism existed in the form of cohesive ethnic groupings and a powerful trade union movement which had fallen under the control of corrupt leaders. In 1910 the Los Angeles reformers selected Hiram Johnson as their candidate for governor. Running against four rival candidates, Johnson won the Republican nomination and then defeated the Democratic nominee in the general election. It will be noted, here again, that the victory of the Lincoln-Roosevelt League was quickly, and, on the whole, easily achieved: within three years a movement, originating in Southern California, had been able to win a smashing victory over one of the most powerful political machines in America.

In many respects, Johnsonism resembled Kearneyism, but Johnson was an abler and far more intelligent demagogue than Kearney. Both movements came out of nowhere; achieved swift and spectacular successes in approximately the same period of time; relied upon the same technique—independent political action; and, once their initial objectives were achieved, vanished with the swiftness that they had come to power. Most of the reforms of the Johnson regime were crowded into the period from 1910 to 1916. Kearneyism, the rise of the Union Labor Party, and the Johnson revolt were alike vitiated by an unprincipled demagogic manipulation of the feeling against Orientals; by and large, the reform leaders of the Johnson era were racists, including Johnson himself, Senator James T. Phelan, V. S. McClatchy, and many others.

The same reasons that compelled Kearney to form a new party forced Johnson to be an "Independent." This was the overwhelming Republican registration and the traditional weakness of the Democratic Party. To keep his political fortunes alive, Johnson was forced to fight the Old Guard within the Republican Party and, at the same time, to develop the Progressive Party as a means of escape should he ever lose control of the Republicans. Nominated on the Progressive Party ticket in 1912 with Theodore Roosevelt, Johnson was able to force Taft into the position of being a write-in candidate. In fact, Taft only polled 3,914 votes in California. Unable to vote for Taft without writing-in his name, many conservatives switched their votes to Wilson so that the Roosevelt-Johnson ticket carried the state by the narrow margin of only 200 votes. Realizing that the independent vote held the balance of power, but that it could seldom win in a three-way fight, Johnson then proceeded in 1916 to conduct a most remarkable campaign for the Senate, in the course of which he conspicuously failed to attack Woodrow Wilson, the Democratic presidential nominee. When the final votes were counted, Johnson had been elected to the Senate by the startling majority of 296,815 votes, whereas Hughes had lost to Wilson by 3,774 votes.

Johnsonism, as a reform movement, expired with Johnson's election to the Senate; but his extraordinary personal career began where the reform movement ended. With each successive election, he was returned to the Senate with ever-larger majorities. In 1928 he polled 75 per cent of the votes cast for the office and, in 1934, 95 per cent. The key to his amazing success always consisted in his complete mastery of California's peculiar style of independent politics based on the weakness of party loyalties and the strong tendency toward independent, non-partisan voting.

The Revolt of the Thirties

In the 1930's California was again ripe for another movement of social protest; the circumstances were, in fact, quite similar to those which had prevailed in 1877. Although Franklin D. Roosevelt had carried the state in 1932, the Democratic Party had failed to capitalize upon the sharp increase in the Democratic registration which Roosevelt's popularity had occasioned. Seeing the opportunity which existed, Upton Sinclair launched his crusade to End Poverty in California one year in advance of the 1934 election. Not more than half a dozen people were involved in the Epic movement at the outset; in fact, the movement was largely Mr. Sinclair's personal conception. Month after month he went up and down the state, forming Epic Clubs, selling thousands of copies of his pamphlet on how poverty could be ended in California. The great farm strikes of 1933–1934, and the San Francisco general strike of 1934, brought the savage repressions which have invariably touched off mass revolts in California. Although functioning within the Democratic Party, the Epic movement was an independent organization and was patterned, by accident rather than design, after the various "clubs" and "locals" of the Workingmen's Party of 1878. Not only did Sinclair win the Democratic nomination for governor, but, had it not been for a Progressive Party candidate, he would probably have been elected. The final vote was Merriam (Republican), 1,138,620; Sinclair (Democrat), 879,557; Haight (Progressive), 302,519.

Sinclairism was Kearneyism and Johnsonism all over again. Once again a mass revolt, in an amazingly brief period, had risen to great power, with little preparation or formal organization. In each case, popular discontent had quickly crystallized in the form of independent political action. Like the Workingmen's Party, the Epic movement swept a number of candidates into office and laid the foundation for important subsequent development; and, like the earlier movement, the Epic organization disappeared almost as quickly as it had emerged. By 1938 there was little left of the original Epic organization. The rise of the pension plan movements of the 1930's follows, of course, very much the same pattern. The repetition, at fairly regular intervals over a period of fifty years, of this same pattern of more or less spontaneous mass revolts, along unorthodox lines, suggests the existence of certain "hidden" springs of action in California politics. One of these "springs" certainly consists in the absence of social organization, of settled forms and channels through which mass protest might find expression. But the absence of these forms does not fully account for the strong tendency toward independent voting in California, i.e. the willingness of the voters, under stress, to form independent political movements for the accomplishment of particular objectives. Although this tendency is made up of a number of factors, it had its origin in a peculiar California issue—the problem of Oriental immigration.

CALIFORNIA'S BLOODY SHIRT

The most striking aspect of California politics in the period from 1870 to 1910 was the overshadowing importance of the agitation against Oriental immigration. In the first three decades after the Civil War, most of the other states were preoccupied with issues which stemmed from the war, notably, pensions, patronage, and reconstruction. But California, which was hardly even a party to the Civil War, was not interested in "the bloody shirt" and other issues which came out of the rebellion. It was, however, greatly interested in the exclusion of Chinese immigration. It was on this one issue that California first began to show a tendency to deviate from national political trends and to assert its political independence.

In 1880 six of seven California electors cast their votes for the Democratic presidential nominee although the state legislature had an overwhelming Republican majority. The reason for this switch was that the Californians were dissatisfied with the position which the Republican Party had taken on the issue of Oriental immigration. In 1884, California was back in the Republican column because the Democrats had not taken a sufficiently extreme position on the same issue. The vote in 1880 and 1884 demonstrated conclusively that the Chinese issue determined the electoral vote of California. And, since the strength of the two major parties was so nearly equal during these years, the Pacific Coast States held the balance of power and largely determined national elections on the basis of a single issue. In these post-bellum years, as Dr. Carl Russell Fish has stated, "independent voting was generally a negligible factor except in the trans-Mississippi district, where the currency moved the farmer, and the Chinese question the Californian, to desert his party standard." Over a period of years, therefore, a tradition of independent voting was built up in California largely in reference to the one issue of Oriental immigration.

Now, how did it happen that Oriental immigration was invested with such massive political weight in California? Dr. Forrest E. LaViolette, in a study of the peculiar behavior of British Columbians in relation to this same issue, has provided the best answer.

The constant uneasiness about Oriental immigration, and the accompanying truculence, suggest to Dr. LaViolette the existence of "profound psychological problems which are an integral part of the struggle to establish a stable society in the Pacific area which has experienced historically a very rapid social change, and, we suppose, consequent social instability." The British Columbians, like the Californians, were quick to realize that the west coast was more accessible to Oriental than to occidental immigration. Geographic isolation, moreover, had sensitized the settlers of both regions to the numbers and kinds of people who were being attracted to the west coast. Since the west coast was obviously destined to experience rapid population growth, the whole question of immigration from the Orient was invested with a particular urgency. Like California, British Columbia needed laborers and sought immigrants; but, precisely because these needs existed, the province was distressed by the importation of Chinese coolies. For if the Chinese came in large numbers, other and more "desirable" immigrants might be discouraged.

Like California, British Columbia is made up of "outsiders": 46 per cent of the population in 1931 was born outside of Canada. Immigrants to an isolated region become highly conscious of their origins, and, in some respects, they come to think of themselves as the defenders of the traditions which they represent. Isolated from their parent groups, they become more chauvinistic than these groups, for they realize that their traditions have been weakened by migration; therefore they feel less secure. The insecurity which isolation breeds makes, also, for a feeling that the remote province is being "neglected," that its problems are not understood, and that it is the victim of discrimination. "Where there is a rapidly expanding population," writes Dr. LaViolette, *"with no established social organization* for distributing economic gain, the struggle for political control, for economic power, and for social prestige soon gives rise to individual feelings of not getting one's share, of having to watch carefully 'the other fellow' and other groups." Paradoxically, it is the peripheral frontier province, not the established and crowded center of population, which is most susceptible to the disease of xenophobia.

In such regions there develops, as Lord Bryce said, "a sort of consciousness of separate existence." New settlers become morbidly conscious of the "destiny" of the region, the place which it will ultimately come to occupy in the larger scheme of things. The importance of the *future* tends to overshadow immediate realities. This self-consciousness in turn focuses "the spotlight of social acceptance and rejection upon groups which are unable to join *quickly* the developing amalgam." Thus the swiftness with which California emerged as a state emphasized the *relative* slowness with which Oriental immigrants, handicapped by racial and cultural differences, adjusted to the new environment. Actually California was so new, so unformed, that there was little in the way of an established social structure or norm of behavior to assist these immigrants in the process of assimilation. The very strenuousness with which the Californians were seeking to forge new social institutions made for a marked irritation with those groups which, on the surface, seemed to be retarding the effort. The "threat" which the presence of Orientals implied was measured in terms of the need to "protect" the infant society.

In a state made up of newcomers, lacking social organization and political traditions, social cohesion can be most quickly achieved by the negative device of rallying opposition to some "menace," real or imagined. Hence anti-Oriental agitation was essentially a device by which the Californians were able to achieve a degree of integration; a means by which they could assert their independence and their "consciousness of a separate existence." One might say, therefore, that it was a social and psychic necessity of the situation. It is only within this large historic frame of reference that one can understand the anti-Oriental movement which, originating in California, ultimately spread around the rim of the entire Pacific basin.

Not understanding this background, the nation was first amused and then amazed when, in 1907, President Theodore Roosevelt invited the Mayor of San Francisco and the members of the school board to the White House to confer, if you please, about an issue involving the foreign policy of the United States. It was as though the President had invited the ambassadors of a sovereign power to meet with him for the purpose of negotiating a settlement of an international dispute; in this case a dispute between "California" and the United States. Nothing quite like this had ever happened before in the history of the nation. But the situation became still more ludicrous when, in 1913, President Wilson sent his Secretary of State to Sacramento to plead with the legislature not to adopt the Alien Land Act. In the prior episode, the Californians had at least gone to Washington; but here the Secretary of State had gone to Sacramento, hat in hand, "to plead with" a legislature as though it represented a sovereign national power. To understand this amazing spectacle, one must realize that the Californians thoroughly enjoyed his newly acquired sense of power; this ability to make Washington jump. Incidentally, the legislature and the governor refused to heed the pleas of the Secretary of State.

HOMEBRED KNIGHT-ERRANT

One cannot get at the dynamics of California politics, however, solely through an analysis of the inter-action of a set of peculiar social and historical forces. Such an analysis is rewarding but it fails to explain certain puzzling aspects of the problem. For example, the Californians have long shown a remarkable disposition to experiment with new social and political ideas. To explain this tendency one needs a sociology of ideas. Without attempting to define the factors which make for this experimental attitude, I would like to suggest the nature of these factors by examining certain phases of the career of that most remarkable Californian, Henry George.

Arriving in San Francisco in 1857, George worked for a time as a farm hand, "sleeping in barns and leading the life of a tramp," and later took part in the Fraser River gold rush. Returning to

California, he had occasion to observe the process of land monopolization at close range as a reporter for a Sacramento newspaper. It was in Sacramento that the first squatter riots had occurred. For two decades, George was an active participant in the bitter fight for land reform in California. His great work, *Progress and Poverty*, was begun the year the Workingmen's Party was formed, and its publication coincided with the calling of the constitutional convention of 1879. *Progress and Poverty*, which Richard Ely once called "the most widely read of modern books on social problems," was a strange kind of book to emerge from a "frontier" community. What was there about the social atmosphere of California that prompted George to write this remarkable book?

"If I have been able to emancipate myself from ideas which have fettered far abler men," he once said, "it is doubtless due to the fact that my study of social problems was in a country like California, where they have been presented with *peculiar directness*." In California George actually witnessed a process of land monopolization which, in one generation, brought about changes that elsewhere had been spaced over centuries. It is reasonable to infer, therefore, that the rapidity with which this process took place produced a sense of shock, of amazement, which greatly stimulated George's curiosity. Above all, it was the appearance of *poverty* in the midst of *progress* that challenged the mind of Henry George. "Amid all the buoyant freedom of a new society," writes his biographer, "there were appearing traces of the symptoms that characterized older and more respectable communities, namely, want and misery and charity." In older societies, these symptoms could be rationalized as the symptoms of social age; but California was new. It was "the disquietude of labor" that most disturbed George and that compelled him to seek an answer to what was then called "the social problem."

Although the ideas in *Progress and Poverty* were not original, there was a remarkable originality and freshness in George's approach to social problems. It was precisely this quality that gave his book its enormous popular appeal, and that makes it so readable today. As Vernon Parrington pointed out, "George was our homebred knight-errant . . . our most original economist." His thinking reflected, as Parrington put it, "the impact of frontier economics upon a mind singularly sensitive to the appeal of social justice." George thought about the origin of things, "as if he were the first man who ever thought" because he was actually an eyewitness to the origin of poverty. It was the "dramatic repetition" of the process of land monopolization in California that so forcefully impressed Henry George. In California he actually saw an economic process "visibly hardening to the static, with a swiftness dramatic enough to impress upon him the significance of a story that had been obscured in earlier telling by the slowness of the denouement." Since this process unfolded in California with such dramatic swiftness, it naturally followed that the remedy which George proposed should have been both drastic and novel. No patent nostrums would serve because the vividness of the experience had thrown into sharp relief his earlier experiences with poverty in the East. "With a fresh young people," wrote his son, "full of self-confidence, and free from restraints and traditions, here were all the conditions needed *to quicken* original thought." The circumstances encouraged originality just as the facts demanded explanation.

The forces that prompted George to make his inquiry into the origin of poverty have continued to operate in California. The paradox of progress *and* poverty is almost as striking in the California of today as in the California of 1879. The richness and abundance of California continue to emphasize the anachronism of poverty. Hence the utopianism of California, its disposition to try perilous remedies for present evils. And, since migration consistently weakens traditional ideas, utopian proposals get a readier and wider hearing in California than in other states. Who has not witnessed miracles in California? The transformation of the desert that was Imperial Valley into one of the great truck gardens of the world, and similar feats, creates an inclination to believe in social and political miracles. Technological inventiveness is matched by a social and political inventiveness. People who have settled in California *know*, as part of their intimate personal experience, that they have been forced to discard many long-accepted ideas and ways of doing things. Therefore they are inclined to believe that social and political ideas which they once accepted as part of the natural order of the universe are not necessarily perfect. They also have a feeling that many of these ideas do not square with the nature of the environment in California. Furthermore the novelty of some of California's perilous remedies for present evils—the single tax, direct legislation, $30-Every-Thursday, and many others —reflect the novelty of the experience which produced these proposals. Such, in a general way, is the social origin of California's utopianism.

THE HIDDEN SPRING

California's progressivism has always consisted of two imperfectly fused but closely inter-related traditions. On the one hand, there is the Kearney-Johnson tradition based upon independent voting, a noisy muck-raking of the "vested interests," a generally pro-labor and public-ownership stand, coupled with a strong racial bias. The other tradition consists of a deeply-rooted indigenous radicalism which has been consistently "socialist" and "utopian," with the emphasis on social and racial equality. Neither Kearneyism nor Johnsonism can be properly understood apart from the dynamics provided by this indigenous radicalism.

If the name of Hiram Johnson symbolizes the progressive tradition in California, that of Burnette G. Haskell symbolizes the older, the radical tradition which has always been the hidden spring of California progressivism. Haskell was born in California in 1857 of pioneer parents, and attended the University of California, the University of Illinois, and Oberlin College. He was admitted to the bar in California in 1879, the year in which *Progress and Poverty* was published and the state adopted a new constitution. He soon abandoned the practice of the law, however, to found a radical labor weekly, *Truth*, which he published in the eighties. A brilliant and versatile man, Haskell was thoroughly versed in the radical and socialist writing of the time and had been particularly influenced by Laurence Gronlund's book *The Cooperative Commonwealth*, which has been called "the first comprehensive work in English on socialism." In the 1880's Haskell formed the International Workingmen's Association, patterned after the International Workingmen's Association which Karl Marx had formed in London in 1864. The IWA was an explicitly socialist movement which, by 1887, claimed 6,000 members in the western states and a membership of 1,800 in California. A dozen or more of Haskell's associates in this movement became outstanding leaders in the trade union movement in San Francisco; in fact, the formation of the IWA had a great deal to do with the revival of the trade union movement in California after its beginnings in the gold rush period. Unlike Kearney, Haskell was opposed to all forms of racism and consistently refused to support the anti-Chinese crusade.

Becoming dissatisfied with trade unionism as a means of social reform, Haskell retired from the movement in 1885 to found a cooperative colony at Kaweah, in the forests of Tulare County. Based on socialist principles, the Kaweah Colony attracted some 400 settlers, most of whom had been connected, in one way or another, with the trade union movement. The colony also included, as Haskell later said, "dress-reform cranks and phonetic spelling fanatics, word purists and vegetarians. It was a mad, mad world, and being so small its madness was more visible." By 1890 the colonists had filed claims on 600 acres of land; had constructed, at great hardship, a model road eighteen miles in length to Camp Gronlund where a town-site had been laid out; owned a sawmill valued at $10,000; and had acquired a site for the development of water power. The property for which the colonists had entered claims was conservatively worth $600,000. The colony boasted the best equipped printing office in the state, published an excellent weekly magazine which had subscribers in nearly every state and in many foreign countries, and had attracted world-wide attention to their socialist experiment. During these early years, however, the colony was under incessant attack, and its enemies finally closed in on the project. On October 1, 1890, an act was passed by Congress which set aside a large area, including the colony lands, as a national forest. The effect of this act, of course, was to make trespassers of colonists who, in perfect good faith, had filed on lands open to settlement. Although a congressional committee recommended that some settlement should be made with the colonists, no action was ever taken on the report. Then the trustees were charged, in the federal courts, with having unlawfully cut *five pine trees* on the public domain and were promptly convicted. Still later, an indictment was returned against them for illegal use of the mails but the judge hearing the case ordered the jury to acquit them.

For many years prior to his death on November 15, 1907, Haskell lived in poverty "friendless and forgotten," a prophet many years in advance of his time. Years after the colony had been forgotten, he wrote these moving and dramatic lines: "I look out of the window of my mountain cabin and the sky is full of storm. . . . Seeing it now, the colony lying dead before me, knowing that its own hands assisted in strangling it, knowing that the guilt of its death rests upon nearly all of its members, myself far from being excepted, the faltering steel that cuts the epitaph chisels as well 'peccavi.' We were not fit to survive and we died. But there is no bribe money in our pockets; and beaten and ragged as we are, we are not ashamed. . . . And is there no remedy, then, for the evils that oppress the poor? And is there no surety that the day is coming when justice and right shall reign on earth? I do not know; but I believe, and I hope, and I trust."

11

THE STATE THAT SWINGS AND SWAYS

◇◇

TURNING FROM the peculiar dynamics of California politics to the contemporary scene, one finds Winston W. Crouch and Dean E. McHenry, authors of a recent book on *California Government*, echoing the judgment of Lord Bryce that "California politics is unique." "Unique," however, is hardly the right word. California is a state that lacks a political gyroscope, a state that swings and sways, spins and turns in accordance with its own peculiar dynamics. The nature of these dynamics has been suggested in the last chapter; the present deals with certain exceptional aspects of the contemporary scene. These aspects have to do with independent voting and the absence of machines and bosses; the shifting sectional pattern of California politics; and the emergence of a new type of lobbyist and new forms of political control.

THE BULGE IN THE MIDDLE

Wendell Willkie once pointed out to John Gunther that the first thing to find out about a state, politically, was the size and temper of its middle or independent vote. According to this formula, the unpredictability of a state varies with the size of its independent vote. When Gunther attempted to apply the formula in California he discovered that "the bulge in the middle is enormous. . . . Only about a third of the people can be counted upon as being on either side; the center is almost as wide as either wing, and almost as variable as the wings are fixed." The early emergence of this "bulge in the middle" accounts for California's liberal election laws, which, in turn, make it possible for the tendency to independent non-partisan voting to find expression. An explanation of these laws, therefore, is essential.

In 1866 California adopted the first primary election law in the United States. Under this law, parties could, if they wished, nominate candidates at a regulated primary election rather than by the convention system. Later, in 1909, the legislature adopted the first direct primary law which ended the convention system. In California, also, judges, county, municipal, and school district officers have, for many years, been elected on the basis of a non-partisan primary. The direct primary law of 1909 contained a rigid test of party loyalty, i.e. the candidate had to declare that he had supported the party at the preceding general election. But Hiram Johnson, fearful that he might lose control of the Republican Party, was instrumental in bringing about a modification of this provision in 1911 and, two years later, the right of a person to become a candidate of more than one party was expressly granted. This, in brief, is the origin of California's liberal nominating system and its notorious cross-filing procedure.

The cross-filing system has made a shambles of party regularity and party discipline in California. In 1940, 55 per cent of the congressional seats were filled at the primary, i.e. by candidates who had captured the nominations of both parties. In 1942 and 1944, 52.17 per cent of the seats were filled in this same manner. The tendency of the primary to supplant the general election is even more marked in the selection of state legislatures. In 1944, 90 per cent of the districts holding elections gave both party nominations to a single senatorial candidate at the primary; and, in the same year, 80 per cent of the candidates for the state assembly were elected at the primary. In 1940 Hiram Johnson was re-elected to the United States Senate at the primary; and, in 1946, Earl Warren was re-elected governor at the primary. For many years the minor state executive offices have been regarded, although they are not so in point of fact, as non-partisan offices. Maine, Maryland, Massachusetts, New York, and Vermont permit, in varying degrees, some form of cross-filing but in no state has this system become so much a part of the political structure as in California.

What the penchant for cross-filing reflects, of course, is the weakness of political parties in California. Party loyalty is a by-product of local tradition operating in settled communities in which social forces are relatively static. In this sense, tradition has a twofold meaning: the traditional Republicanism of particular localities, states, and regions which finds its counterpart in the Solid South's adherence to the Democratic Party; and tradition as an aspect of family loyalty. Party organization is based on political regularity and political regularity is based on tradition. Before tradition can become a factor in either sense, however, it is essential that the voting population should be relatively static. A glance at a map will show that the areas in the United States which have been most consistently Republican or Democratic are areas which have received little migration. They may be, and often are, areas which export population; but, by and large, they are areas in which population growth is based on natural increase. Political traditions are, by their nature, localized and personal.

Based on a sense of party regularity, political machines function most effectively in neighborhoods and communities in which the various social elements have achieved a degree of equilibrium, and in which the leaders of these elements are well-known and clearly identified. In such communities, the recipients of political favors can be easily rounded up and carted to the polls on election day. The basic unit of most political machines is the ward and the ward, in social terms, means an area of settled residence. The strength of the ward leader or "boss" is that he is known to everyone and, in turn, knows everyone. Ward politics are face-to-face politics, based on propinquity, family ties, favors, obligations, alliances, and patronage.

Merely to sketch the conditions of machine politics is to explain why party regularity is a negligible factor in California politics. Migration is the key to independent non-partisan voting in California. Party regularity has been most pronounced in California in precisely those areas which have been least affected by heavy inmigration. Political machines simply cannot function with efficiency in areas which are largely made up of newcomers and strangers. At each presidential election from 1900 to 1948, between one-tenth and one-third of the registered voters in California have cast their first ballots in the state. These "new votes" have been a constantly disturbing factor in state politics and, also, a most unpredictable factor. No sooner are the "new votes" of one election assimilated to the environment and made subject to local pressures and influences, than another wave of new voters has entered the state. With internal migration being almost as tumultuous as migration into the state, it has been impossible to build political organizations that can survive from one election to the next. There are neighborhoods in Los Angeles in which it would be quite safe to say that not more than two or three families in an entire block are known to each other. Try to boss such a neighborhood! In such areas it is extremely difficult even to effect a loose skeleton-like organization for a particular election and it is a foregone conclusion that the effort will have to be repeated, entirely from scratch, at the next election.

In 1948 a study made in San Francisco—far more settled, in this respect, than Los Angeles—indicated that 79 per cent of the voters did not know the names of the candidates for the state legislature in the districts in which they lived and that 67 per cent did not know who was running for Congress in their district. In Los Angeles, 14 per cent of the voters were stricken from the rolls for failure to vote in the 1948 election, and it is a safe surmise that many of these voters had failed to vote because they had changed their place of residence. The impact of migration on California politics is also seen in the remarkable shift in party registrations. In 1924 the Democratic Party had only 22 per cent of the total registration by comparison with a 65 per cent Republican registration (2.5 per cent were registered in minor parties and 10.5 per cent declined to state their party affiliation); but in 1945 the Democrats had 58 per cent of the total registration and the Republicans had dropped to 37 per cent. To some extent this shift was due to changes in registration based on conviction or preference, but the basic explanation is to be found in the fact that the point of origin of migration had shifted from Republican to Democratic areas.

From the earliest date, California voters have been notably independent. For a long time, the state had the highest percentage of "decline-to-state" registrations to be found in any state. Since the days of the Southern Pacific machine, and even earlier, California voters have distrusted political parties as such; witness the adoption of the first primary law in 1866. Studying state elections for the decade from 1920 to 1930, Gosnell could find little correlation between the size of the vote for the office of governor and the vote for United States Senator. A Republican candidate for one office would receive a heavy vote which could not be correlated with the size of the vote for the Republican senatorial nominee and vice versa. "Voters in California," Gosnell concluded, "changed their party allegiances from election to election in accordance with changing issues and personalities rather than adhered to any party symbol." To a large degree, of course, this fluctuation in the size of the partisan vote reflects the almost complete breakdown of party regularity and party discipline but there is another factor involved. Dewey Anderson has suggested that party allegiances exist in California but that they are based on national rather than on state politics. In other words, California, as a state, is "different." It has special issues and peculiar problems which constantly disrupt party allegiances in state elections. So far as national elections are concerned, California is one of the states which, as Gosnell says, "swings with the nation" but swings more violently than other states *unless*, as in the famous Johnson-Hughes case, national issues happen to cut across peculiar local issues.

In studying California's preference for independent non-partisan voting, the absence of party discipline is extremely important. Under the state's free-wheeling style of politics, candidates must depend upon individual political merchandising, that is, they must "sell" themselves as candidates. Since they cannot rely upon a heavy backlog of partisan support, candidates are inclined to thumb their noses at the local and state officials of the party in which they are registered. The ambition of every candidate in California is to win at the primary; hence they must maintain a general aura of non-partisanship, however partisan their voting records may be. With the trend toward independent voting being so strong, most candidates are extremely reluctant to be identified with a ticket or slate unless the head of the ticket happens to enjoy great personal popularity. Candidates think in terms of personal machines, personal followings, individual campaigns, and not in terms of party organization. Party discipline has declined, in fact, in direct ratio to the increasing emphasis placed on cross-filing.

Party discipline and regularity has also been undermined, to a degree, by the initiative, referendum, and recall which have been available to the voters of California, as a means of direct legislation, since October 10, 1911. As a matter of fact, the recall originated in Los Angeles where it was adopted, as part of the city charter, in 1903. California's fondness for these devices reflects, again, its distrust of political parties and machines. Although the recall has been sparingly used, the initiative and referendum are favorite devices. At almost every general election, the ballot will contain from twenty to thirty special initiative and referendum issues. Apart from other consequences, the existence of these forms of direct legislation has had a tendency to weaken party discipline and regularity.

If the breakdown of party responsibility has had some highly unfortunate consequences in California, it must also be said that it has had certain advantages. With the exception of the William Malone machine in the Democratic Party, in San Francisco—essentially a local "operation"—there are no political machines, as such, in California. Theoreticians attribute the absence of political machines after the fashion of the Hague, Kelly-Nash, and Pendergast machines, to the existence of such measures as the direct primary, the initiative, referendum and recall, and the system of cross-filing. Actually the peculiar social structure of the state is the basic explanation. Whatever the reason, however, California voters will be extremely reluctant to abolish the system of cross-filing if only because it has minimized blind party regularity and machine politics.

Although migration is the key to the prevalence of independent non-partisan voting in California, there are, of course, other factors involved. "The independent tradition," writes Dean McHenry, "born of the 1910 revolt against the Southern Pacific machine, carries a suspicion of parties and a reliance upon individual leaders." Then, too, the fact that many offices are conducted on a non-partisan basis has made it extremely difficult to build political machines or to develop much interest in party activities. Merit systems—state and local—are so widespread that there is not much room for patronage politics. Frightened by the prospect that Upton Sinclair might be elected governor in 1934, the legislature "blanketed" virtually all state positions under civil service so that today, apart from replacements and appointments to new offices, the governor has only about 59 appointments, out of 30,000 state positions. In other words, a governor finds it extremely difficult to play patronage politics in California and, like the candidates for the state legislature, he must rely upon individual political merchandising rather than on party loyalty or patronage. Hiram Johnson and Earl Warren are both products of this peculiar style of politics in which "independence" and "non-partisanship" are carefully stressed.

Just as the legislature will not stand hitched in Sacramento, so the California congressional delegation is, by all odds, the least disciplined of any state delegation in Congress. "California," writes Warren B. Francis, "has the dubious distinction of being represented by individuals who most often vote contrary to the principles laid down by both Republican and Democratic chieftains." In 1948 two California congressmen had the "worst" records in Congress on the score of party regularity. If anything, the Democrats are more variable than the Republicans. Customarily the degree of harmony between voting records and party commitments varies from 44 per cent to 95 per cent in the California delegation.

With this remarkable record of independent non-partisan voting, based upon the most "liberal" and "progressive" election laws of any state, and the absence of political machines, one might assume that California politics were exceptionally free of corruption, graft, and dictation. But, paradoxical as the statement may sound, the California legislature is perhaps more completely under the thumb of one individual than any state legislature. Before attempting to explain this striking paradox, however, it is necessary to examine the amazing career of Artie Samish, the King of the Lobby. In one of his talkative moods, Artie once said: "I am the governor of the legislature; to hell with the Governor of California." Boss of the legislature since 1932, it would be extremely difficult to find, in the entire range of American politics, a more extraordinary political virtuoso than Artie Samish.

THE GUY WHO GETS THINGS DONE

The names of many Californians are listed in the index of *Inside U.S.A.* but one name—politically the most important name—does not appear: the name of Artie Samish. Although Samish is known to everyone in California who is directly interested in politics, I would venture the guess that not one per cent of the voters could identify his name although he is, beyond all doubt, the political boss of the state. Nationally he enjoys an even more remarkable degree of anonymity. What, then, is the basis of the amazing power of this anonymous boss who functions in a state in which theoretically there should be no bosses and in which, in point of fact, there are no old-style political machines? The answer is to be found in the peculiar nature of California as an economic empire but, first, a word about the background of Artie's career.

From 1880 to 1910, California was "bossed" by the Southern Pacific machine. In those days, S. T. Gage, of the Southern Pacific, was widely known as "the King of the Lobby." During this period there was, in effect, only one lobby, that of the Southern Pacific, and the function of the other lobbyists was to lobby this lobby. The vulnerability of the company, however, consisted in its notoriety: everyone knew that the Southern Pacific controlled the legislature. For a brief period, from 1910 to 1916 (under Hiram Johnson), California managed to extricate itself from the control of the machine-lobby and the state's indigenous liberalism found immediate expression in one of the most remarkable records of reform legislation ever adopted in a single state in a comparable period. It is significant that the power of the Southern Pacific machine-lobby prevailed during the period in which there was, in effect, only one political party in California, namely, the Republican Party. Reform came when Johnson challenged the Old Guard in the Republican Party, and in doing so, made it possible for the people to challenge the machine.

But, by 1920, a new lobby began to make its appearance. It included labor, agriculture, the women's clubs, teachers, a vast range of business interests, reform groups, and whatnot, all jockeying for power and undercutting each other. With party discipline at a minimum, a vacuum had been created at Sacramento which had to be filled; if neither the governor nor the political party chieftains could boss the legislature, someone had to undertake this function and that someone was Artie Samish who has referred to himself, quite accurately, as "the guy who gets things done."

Not too much is known about Artie's personal background as he has had until recently appreciated the importance of keeping his name out of the headlines. At times this has been an extraordinarily difficult personal performance for Artie is extremely vain and would like it known that he is the boss of the state but, when temptation has arisen, he has always remembered that notoriety caused the defeat of the Southern Pacific machine. Samish started out as $170-a-month clerk in the tax collector's office in San Francisco. In about 1918 he got a job as minute clerk in the legislature and in this position acquired a complete mastery of the mechanics of legislation. No one in California today has a more expert knowledge of the business of legislation than Mr. Samish, the ex-minute clerk whose annual income, as a lobbyist, has for many years been in excess of $200,000.

The thirties, of course, was an ideal period for the professional lobbyist. Liquor was coming back and with it gambling and horse-racing, interests which have always spent money lavishly in politics. More important than these interests, however, was the drive of business for control measures aimed at fixing prices and eliminating "unfair" competition and "unfair" trade practices after the manner of the NIRA codes. Various professional associations, also, were concerned with eliminating potential competitors through the guise of stricter standards and a greater emphasis on "professional ethics." Trade associations, concerned with "protection" in one form or another, multiplied; 46 trade associations had registered lobbyists in Sacramento in 1933. All in all, 339 lobbyists were registered in Sacramento that year, three lobbyists for every legislator. It was in relation to this background that Artie scored his first major triumph in 1931: an amendment to the Public Utilities Act known as "the famous section 50¼" which gave marked competitive advantages to the Pacific Greyhound Bus Line which is owned by the Southern Pacific. From 1935 to 1938, Samish received $61,237.48 from the Motor Carriers Association which he had organized to eliminate "wild-cat" trucking operations.

In the wake of repeal, Samish lobbied through the Alcoholic Beverage Control Act—for a fabulous fee. From 1935 to 1938, Artie's annual fee from the California State Brewers Institute was $30,000. Through Frank X. Flynn, one of his lieutenants, he also represents the somewhat conflicting interests of the California Liquor Industries Association ($12,500 for the same three-year period); the Wholesale Liquor Dealers Association of Northern California ($12,000); and the Wholesale Liquor Dealers Association of Southern California ($24,500). In addition, the brewery interests levied an assessment of 5¢ a barrel which went into a fund which was turned over to Samish to spend as he saw fit for general political activity, without any formal accounting or records. This fund, which totaled $97,619.47 for the three-year period in question, is still maintained.

Then, of course, there was the matter of special legislation for the Los Angeles Turf Club, which operates the Santa Anita race track. Between 1935 and 1938, Artie collected $54,999.97 from this source. Subsequent to 1931, other rich plums fell into his lap with the result that the Philbrick Report, an official state document, states that $496,138.62, in addition to the special assessment of $97,619.47 from the brewers, passed through his bank account in the period from 1935 to 1938.

The pattern of Artie's operations is quite clear. When asked to represent a special interest group, his first step is to organize a trade association, if one is not already in existence, and to secure a contract as its "public relations counsellor." In most cases, the trade association is or rapidly becomes his alter ego. "There is no difference," he once said, "between Arthur Samish and the Motor Carriers Association. I *am* the Motor Carriers Association." Once he has a contract with the trade association, he will secure the legislation in which the association is interested, or repeal or amend "unfavorable" legislation. But, at this point, the trade association discovers that what Artie has given, Artie can take away; hence the annual retainer continues indefinitely. If the association or the individual client becomes restive, a reminder is usually forthcoming that Mr. Samish is a good man to retain. For example, the American Potash and Chemical Company, which retained Artie in 1935 to lobby against a proposed severance tax discovered in 1937 that mysterious news items and full-page ads discussing the likelihood of a renewed campaign for the severance tax were appearing in the press. "Mr. Samish," as the Philbrick Report cryptically comments, "forwarded these items to the American Potash and Chemical Corporation," and that year his retainer was advanced from $6,500 to $14,000. Diligent in behalf of the interests he represents, Mr. Samish is always quick to pick up items of this sort, and his clients, after thinking the matter over carefully, interpret these items correctly.

It is Mr. Samish's distinction that he has invented a new type of machine carefully designed to meet the peculiar requirements of California politics. He is, in other words, a new type of political boss. In the past, most political machines have existed through their control of the party apparatus; the party machine controlled the legislature. But in California there are no party machines; in fact, it is almost true to say that there are no parties. What Mr. Samish has done, therefore, is to convert the interest-group into a political machine which functions independently of the party. From the lobbyists' point of view, of course, this represents a distinct advance in the forms of political control. A party machine can be challenged at the polls but as long as Artie controls the interest-groups his power is beyond dispute. Theoretically his power could be challenged by the interest-groups he represents but—and this is the key to the structure of power he has fashioned—these groups enjoy, despite the costs, great advantages from his representation. In the first place, the state takes over the function and also the expense of policing the particular industry against "unfair trade practices"—an enormous saving in itself. In the second place, each industry-group and each individual member of this group is spared the trouble and expense of dealing with individual politicians. Control must be centralized to be effective and, in California where there are no old style party bosses and little party discipline, business and industry must have some "protection" against the endless demands of free-lance politicians.

Quite apart from the power which he derives from his annual fees and special commissions, Mr. Samish has not one but a dozen different machines at his command. For example, in 1938 there were 50,000 retail liquor outlets in California (there are more today). The owners of these outlets are all political legmen for Mr. Samish; in effect, they are the "ward bosses" of this new-type machine. They are distributed, moreover, throughout the state and in relation to the density of population. It is certainly to be doubted if Pendergast himself ever had as many lieutenants at his command. But Artie has many more because the liquor dealers are not his only clients. In short, he uses the trade association as an old-style boss would use a patronage machine. But patronage machines have a tendency to get out of hand—there are always minor rebellions and incidents of careerism on the part of lieutenants to cope with. Artie's machine, on the other hand, functions with matchless efficiency, and is self-regulating. He has no lieutenants; only two "assistants"—one for the North, and one for the South. Most of the trade associations that he represents publish bulletins or journals, and he controls these publications. The trade knows what he thinks it is desirable that it should know—and no more. Furthermore, control of the brewers' five-cents-a-barrel fund gives him an enormous campaign fund which is all the more effective since there are virtually no political chieftains or party bosses to deal with.

From 1931 to the present time Samish has controlled a large bloc of votes in the state legislature. Control of this bloc is tantamount to control of the legislature, as this bloc usually elects the speaker of the Assembly, who appoints the committees. Majority control of two or three key committees carries with it, of course, the power to kill in the committee or send out with a "do-pass" recommendation most important pieces of legislation. As the Philbrick Report revealed, Samish employs individual lawyer-legislators as "counsel" for the various trade associations that he represents. In fact, it is his boast that he employs more lawyers than anyone else in California. But this is merely one facet of his power over legislators. For in the absence of party machines, given the cross-filing system and the trade association machines which he controls, he can nominate and elect candidates in many districts by the expenditure of nominal sums. Furthermore his power cannot be exposed since the most powerful economic interest-groups of the state are his clients. For example, he keeps up-to-the-minute files on the allocation of advertising by his various clients. If a newspaper is "unfriendly," advertising is promptly removed.

It should be noted, also, that Artie is quite "fair" with his stable of legislators. He doesn't care a rap what they vote for or against, so long as it does not affect his clients. Most of the legislative items in which he is interested, moreover, appear to be quite innocuous and routine matters. They often have to do with changing three or four words in a certain section of a most involved statute; but by one such amendment he kept an eastern distillery from doing business in California. Voters do not pay much attention to these items although thousands of dollars may be involved; in most cases they never hear of the legislation. It is not surprising, therefore, that some of Southern California's "liberal" and "progressive" legislators have been in Artie's stable for years without their constituents having the slightest knowledge that they are "Samish men."

Above every other consideration, it is Samish's ability to sell the police power of the state that accounts for his extraordinary power. California has been one of the most active states in regulating general business by statute. In 1907 it followed the federal government's lead and adopted a state anti-trust act. Two years later the statute was amended to provide that no agreements and combinations were illegal if the purpose was to ensure a reasonable profit. "The State Supreme Court interpreted this in a manner," as Dean McHenry has pointed out, "to permit manufacturers to set a price for their commodities and to require retailers to observe the figure." Then, in 1931—the year of Artie's emergence as a big-time lobbyist—the legislature adopted the Fair Trade Act which put these previous court rulings into statute law, and the act was further strengthened in 1933 by permitting firms which suffered from price-cutting activities on a fixed-price item to sue the offenders.

To understand the importance of various "control" measures in California, it should be kept in mind that constant migration has brought to the state thousands of people who, in their eagerness to get a foothold in business or the professions and to overcome the disadvantage of being newcomers in highly competitive fields, have frequently cut prices, engaged in "unfair" trade practices, and demoralized more than one "gentlemen's agreement."

Little effort has been made under this act to enforce competition; on the contrary, the drive has been to eliminate "cutthroat" competition; "wild-catting," and the like. A retail liquor dealer may violate one or more of the countless regulations of the ABC Act and escape punishment. In fact, these regulations were adopted so as to make it practically impossible for a dealer to operate a liquor store without violating the law (the better to put the heat on in reference to more important matters). But let this dealer cut the price of a bottle of bourbon by so much as a nickel and his establishment will be swarming with state inspectors. Businessmen, not ward-heelers, are the lieutenants in the Samish system. Artie does not have to employ "goons" to police an industry; the state does the policing and the public pays the bill. If any individual businessman, a member of an industry group represented by Samish, feels a disposition to rebel against this system, he will soon discover that he is fighting, not Samish, but the State of California. There is nothing furtive or undercover or devious about this system; it is all open and above board and perfectly legitimate as measured by the current code of business ethics. All that has happened is that the police power of the state has been sold to private interests.

CAMPAIGNS, INC.

In tracing certain phases of the political career of Arthur Samish, I may appear to have lost sight of his colleague of former years, Clem Whitaker; but Clem really deserves a separate section. From being a lobbyist so powerful that, on one occasion, he sat at the elbow of a state senator, on the floor of the senate, and told this distinguished legislator what to say in the course of a debate,

Mr. Whitaker has come to specialize in initiative campaigns. He has, in effect, a working jurisdictional agreement with Mr. Samish: Artie handles the legislature, Clem takes care of the initiative campaigns. The division is logical, for control of the legislature would be ineffective unless some instrumentality existed to keep the people from adopting all sorts of "crazy legislation" by means of the initiative. Such an instrumentality exists in California; it is known as Campaigns, Inc., and is controlled by Mr. Whitaker.

Campaigns, Inc., which has offices in San Francisco and Los Angeles, is in the business of organizing campaigns both for and against initiative and referendum measures. In any particular election, Campaigns, Inc. usually runs half a dozen separate campaigns. In the last fifteen years, the concern has handled 65 separate campaigns, usually running five or six campaigns at the same time; and it has a fabulously successful record. In 1948 the company defeated Proposition No. 13, a reapportionment measure, and carried Proposition No. 3, an "anti-featherbedding" measure. Mr. Whitaker also runs the California Feature Service which pumps all sorts of material, editorial and otherwise, into the rural newspapers of the state. Needless to say, Campaigns, Inc. represents scientific precision politics. Experience gained in one campaign becomes part of the know-how of the organization in handling other campaigns. There is not a trick in the trade that this organization has not mastered, and it knows how to get the most for its money in the way of publicity and promotion. Its slogans are works of art, and its manipulation of public opinion is something to excite wonder and amazement. It is also, of course, a fabulously successful business. By running so many campaigns over such a long period of years, Campaigns, Inc. has built up a network of friendly alliances, contacts, and feeders-of-information; nor should it be forgotten that the concern controls an immense amount of patronage in the form of printing contracts, advertising, billboards, and so forth. Twice successful in defeating a health insurance program sponsored by Governor Earl Warren, Campaigns, Inc. has now been retained by the American Medical Association for a fee which is said to be in the neighborhood of a million dollars, to defeat President Truman's compulsory health insurance program. In 1942, however,

Mr. Whitaker managed Earl Warren's successful campaign for governor.

Campaigns, Inc., of course, merely defeats or enacts initiative and referendum measures. It is "the doctor" consulted by politically worried patients. But there is a similar organization which specializes in qualifying initiative and referendum measures, namely, Robinson & Company of San Francisco. This company was formed by Joseph Robinson shortly after the first World War. "I was looking for a business with no competition," says Mr. Robinson, "and I found it. We are the only firm of our kind in the country." In the last thirty years, Robinson & Company has qualified 98 per cent of all the measures which appear as special "Initiative Proposals" in California. Mr. Robinson, rather like Mr. Samish and Mr. Whitaker, regards his business as a "semi-public institution." He does not discriminate against initiatives on the basis of their contents. In 1948 it took a minimum of 204,762 valid signatures to qualify an initiative proposal. This meant that to be on the safe side, 250,000 or more signatures had to be obtained. Mr. Robinson's usual fee—he operates on a "we-deliver-or-your-money-back basis"—for qualifying an initiative proposal is $75,000. In 1948 he qualified five out of nine initiative measures on the ballot. For a fee of $180,000 he is also prepared to notify 5,062,089 registered voters in California that a measure on the ballot should be passed or defeated.

The more one ponders such feats of organizational skill as those brought off by Samish, Whitaker, and Robinson, the more one is inclined to agree that there is, indeed, "something peculiar about California politics."

SACRAMENTO: COMMODITY MARKET

Reverting now to Mr. Samish, the real secret of his astonishing political power is to be found in the fact that California is not one thing economically but everything. Somewhat larger than the British Isles, it is in many respects more richly endowed with natural resources. Only New York, of American states, can possibly rival California in the *diversity* of its economic interests. It was precisely the diversity of these interests which enabled the Southern Pacific to boss the state politically for nearly half a century. Where there are *many* interests to be served, there is always a competition for favors. Where a single interest is dominant in a state, as say, "copper" in Montana, or "divorces" in Nevada, the possibilities of political merchandising are limited; there are few opportunities for "deals" and "trades" and "cinch bills."

John Gunther had this consideration in mind when he wrote in *Inside U.S.A.* that "the United States is an enormously diversified nation, and a legitimate lobby can usefully fill a special role, that of representing groups which otherwise have no special representation." Now in terms of diversified economic interests, California is to the other states what the United States is to other nations. "Agriculture" in California is not "wheat" or "corn" or "cotton"; it is 214 highly specialized crops. In fact, as Dean McHenry has noted, the legislative needs of agricultural producers in the state are so varied and specialized that "common problems do not apply to any large proportion of farms." The principal "common problem" is labor. "Banking" is not one thing in California: for a quarter of a century it has meant a bitter fight between one bank, the Bank of America, and the other banks of the state. "Medicine" does not mean doctors and dentists: it means osteopaths, chiropractors, naturopaths, Chinese herb doctors, psychoanalysts, and others. Although many of these interest groups have a broad similarity of interests in relation to certain issues, there are also a great many differences. Hence the fantastic "angling," trading, and jockeying for position that makes of Sacramento one of the great commodity markets in America where an astonishing variety of interests bid for favor and preference.

To ride herd on all these interests, "a governor of the legislature" had to be chosen. If Mr. Samish had not filled this post, someone else would have done so. "The Third House" in Sacramento, made up of lobbyists, who outnumber legislators three to one, has long since been institutionalized and placed on a par with the other two houses. This institution is known as the Pal's Club and is made up of the wives of lobbyists *and* legislators. The name stands for Protective Association of Lonesome Souls, or "Pals" to you. It is highly fitting and proper, therefore, that the social climax to each session of the legislature should be, not the Governor's Inaugural Ball which opens each administration, but Mr. Samish's fabulous parties given at the close of each session of the legislature. Since all the diverse economic interests cannot be represented at Sacramento, and since some must be given the right-of-way in order to get anything done, an auction had to be devised at which these interests could bid for preference. Sacramento is that auction and Mr. Samish is the auctioneer.

How does it happen, it will be asked, that all of California's economic interests cannot be fully represented in the state legislature? Since 1926 less than 6 per cent of the voters of California have elected a majority of the state's 40 senators. Los Angeles County with 39 per cent of the population—3,584,000—has one state senator and so does the El Dorado–Alpine–Amador senatorial district which includes 24,920 residents. The City and County of San Francisco with a population of 750,000 has one senator, and is thus on a par with Mono–Inyo which has a population of 12,270. The vote of 5 state senators, representing 6 million Californians, can be offset in Sacramento by the votes of 5 senators representing 150,000 constituents. Yet the 5 counties with 6 million population pay 69 per cent of the sales tax which is the state's principal source of revenue.

This system of representation has created the political vacuum in Sacramento which is filled nowadays by the portly and influential presence of Mr. Samish. The political demands of the constituents of a "cow county" are rather primitive: a new bridge, the extension of a highway, some minor patronage, a favor here and there. Not only can a senator from one of these counties be reelected at a minimum expense—$5,000 will run an effective campaign—but he is under far less pressure than a senator from a populous county. The voters in the "landscape" counties are not particularly concerned with how their representative votes on state-wide issues which do not affect them. If he gets the bridge, he can vote as he likes on race track legislation. A few thousand dollars in campaign contributions spent in six or eight key districts can produce amazing results. How did it ever happen, therefore, that the progressive and independent voters of California could have acquiesced in such an undemocratic scheme of representation? More important, why are they seemingly powerless to change it?

Until the "boom of the eighties," Los Angeles was known as the "Queen of the Cow Counties," a sleepy province in the south, long dominated by the populous mining districts of the North. But in the decade from 1880 to 1890, the southern counties showed a rate of population increase which was 8 times the average for the state, and in the decade from 1900 to 1910, the South's rate of increase was 4 times greater than the average for the state. Then came the spectacular upsurge in population in the twenties when 2,000,000 people moved into California, two-thirds of whom settled in southern California, with Los Angeles county recording a gain of 1,272,037. The increase in the South's population for this decade was even more spectacular than these figures indicate for most of it took place in the early years of the decade. Within the southern California counties, moreover, most of the increase in population from 1880 to 1930 took place in one county: Los Angeles. This sharply differential rate of growth naturally had profound repercussions in the northern and central portions of the state. Should this rate of increase continue, so these sections reasoned, it would be merely a question of time until one county dominated the state legislature.

The reapportionment issue first assumed major importance in 1911. At this time San Francisco had 17 per cent of the state's population (by comparison with 27 per cent in 1880), while Los Angeles had increased its percentage from 4 to 21 per cent. The conflict over representation, also, reflected the growing trade rivalry between San Francisco and Los Angeles. More basic than this rivalry, however, was a sharp rural versus urban cleavage. This cleavage was more pronounced in California than in most states since, from an early date, the population of the state has been highly urbanized. In 1870 thirty-three per cent of the population lived in urban areas; in 1900, 52.4 per cent; in 1940, 71 per cent. At the time of the 1911 debate on representation half of the state's population had come to be concentrated in three counties: San Francisco, Los Angeles, and Alameda. Not only were the rural areas disturbed by this top-heavy urban representation, but they were particularly concerned by the character of San Francisco's delegation in the state legislature. San Francisco, in those years, was the "home of the machine" and its delegation was suspected of corruption. Agriculture was still the chief income-producer in the state in 1911, and the agricultural counties were naturally concerned over the influence of labor in the three major urban districts. At this time, however, the urban districts held together, and a move to cut down the representation of the cities was defeated.

But, with the great increase in population in Los Angeles in the twenties, San Francisco and Oakland began to share the uneasiness of the rural counties. Various schemes of reapportionment were debated but the legislature was hopelessly deadlocked. To break this deadlock, a so-called "federal plan" was submitted to the voters as an initiative measure in 1926. Under this plan no county could have more than one state senator and not more than three counties might be combined to form a single senatorial district. To the surprise of many people, the measure carried in every county in the state with the exception of Los Angeles. Obviously rural jealousy had combined with the envy of San Francisco and Oakland to change the balance of power within the state. In November, 1928, the measure was resubmitted to the voters and upheld by 61 per cent of the votes cast.

Although the "federal plan" shifted control of the senate from urban to rural areas, the lobbyists, not the rural people, have profited by the change. In 1926 the Los Angeles Chamber of Commerce opposed the "federal plan" but in 1928 it actively supported the measure, for the business interests had come to realize that a lack of democracy has its advantages to special interests. On a long list of measures favorable to the rural areas, California's rural-dominated state senate has voted *against* the interests of these areas. To date, however, the rural voters have failed to see anything suspicious in the circumstance that urban industry, on a state-wide basis, has become the strongest partisan of the "federal plan."

In 1948 the State Federation of Labor sponsored an initiative measure—Proposition No. 13—which would have reapportioned the state senate by redistributing senatorial representation. The measure did not propose an increase in the number of senators, fixed by law at 40, but would have allocated 21 senators to the four major urban areas. It was essentially a compromise measure aimed at striking a balance between urban and rural areas; between representation-by-territory and representation-by-population. With 6 million Californians being represented by 5 senators and 4 million by 35 senators; it could hardly be denied, so the proponents argued, that some readjustment was necessary. On the face of things, moreover, it seemed likely that the reapportionment measure would carry since the five counties most discriminated against had 61.8 per cent of the population. At the time the measure was proposed, however, I wrote a piece for *The Nation* in which I predicted that the measure would encounter much greater opposition than might be anticipated. The prediction proved to be an understatement.

Almost before the measure had been drafted, the State Chamber of Commerce and its affiliates, north and south, were on record against it. The farm groups also opposed the measure. Without exception, the dominant economic groups, regardless of sectional divisions, or farm or city background, took a strong stand against reapportionment. The state association of boards of supervisors unanimously condemned the proposal. Governor Earl Warren came out against it, as did the Lieutenant-Governor, who happens to be from Los Angeles. The opponents of the measure spent $294,772 to bring about its defeat and southern California contributed nearly half of this sum. San Francisco contributed the balance. In other words, the central portions of the state and the rural areas did not foot the bill, although the measure was supposed to be of special concern to these areas; on the contrary, the money came from the urban areas which would have benefited from its adoption. When the votes were counted, it appeared that the measure had lost by a vote of 1,069,899 for, 2,250,937 against. Not a single county in the state supported the measure. This amazing spectacle of a people approving their own disfranchisement can only be explained by the control which the present system of representation gives to the dominant economic interests.

From 1926, when the federal plan was adopted, to the present time, the lobbyists have controlled the state legislature. A governor can bargain with the lobbyists, since he still retains the veto power; but Arthur Samish can, if necessary, block any legislation in the state senate. By and large, the governors of the state have hesitated to challenge the power of the lobbyists, for they realize that behind a man like Samish is a formidable set of interests. Any governor who chose to take the issue to the people would immediately discover that he was fighting, not Artie Samish alone, but also his clients who represent the most powerful alliance of professional, business, agricultural, and industrial interests.

As a result of the failure of the people to get rid of the federal plan, Sacramento has become, not so much the capital of a great state, but the headquarters of the lobbyists. A California legislator got at the real issue when he told John Gunther that "the lobby is the American substitute for the one good thing that distinguished the corporative state, namely, the direct representation of interest-groups." Interests, not people, are represented in Sacramento. Sacramento is the market place of California where grape growers and sardine fishermen, morticians and osteopaths bid for allotments of state power. Today there is scarcely an interest-group that has failed to secure some form of special legislation safeguarding its particular interests. State power, in short, has been pre-empted by special interests. And as Dean McHenry has pointed out, the Sacramento system is "dangerously suggestive of Mussolini's notions of the corporate state." Decisive power rests not with the people but with the Canners' League, the Wine Institute, and the other organized trade, crop and industry organizations. The California legislature, indeed, closely resembles the "chamber of corporations" under Italian fascism. Curiously, the legislators have failed to catch the meaning of the new dispensation; that is, they have failed to form their own trade association and retain a lobbyist. Their salary is still $1,200 a year. However, as the Philbrick Report revealed, most of them manage to do very well.*

---

* Since this chapter was written, the appearance of my article "The Guy Who Gets Things Done" in *The Nation* of July 9, 1949, and two excellent articles by Lester Velie in *Collier's*, "The Secret Boss of California" August 13 and 20, 1949, have served to acquaint Californians with their unknown boss.

12

INDUSTRY: THE CULTURAL
APPROACH

◇◇

FOR THE past eight years California chambers of commerce have been on a glorious promotional binge. Statistics and reports have rolled from the presses and mimeograph machines in endless quantities, boasting of new "peaks" in industrial expansion, in commercial and residential construction, in the growth of western financial power, in the extension of markets. Chronic pessimists have been caught up in the exuberance of the statistical optimists and even the optimists have been surprised to see their not-too-firmly-believed-in predictions of yesteryear topped by the figures and reports of the present.

The more one studies these reports, however, the more difficult becomes the task of trying to summarize their contents. One can analyze the growth of California industry from many points of view: as a regional phenomenon; in relation to national trends; in historical perspective; as a by-product of the war; and so forth. But the more one mediates between these different points of view, the more one is impressed by, the fact that certain underlying dynamics have long been at work in the growth of California industry. Hence I have elected to deal with California industry as a cultural phenomenon. This is obviously a special and limited point of view but it may bring out certain neglected and imperfectly understood aspects of the growth of industry in the state.

I should point out, perhaps, just what is meant by a "cultural approach" to the growth of industry. Essentially I mean this: that the growth of industry in California has always been influenced by certain exceptional climatic, physiographic, and geographic factors which have made for a departure from the national norm. Too frequently the growth of industry has been treated as though industry merely represented a bookkeeping transaction in which the sole determinants were manpower, sources of energy, raw materials, transportation facilities, markets, and capital. Studies of this sort have carried the implication that the rate of industrial growth represents a kind of calculus based upon these and other factors. But there are other factors, less specific and influential, which also bear upon the growth of industry: the accidents of history, the peculiarities of geographical position, the rate of technological development, and, above all, the way in which new human needs emerge in a society and are met or satisfied, and the intangible impact of mind on mind under novel conditions. It is from some such a point of view that I have tried, in this chapter, to examine the growth of industry in the state.

THE MOMENTUM OF AN EARLY START

Studies of industrial growth have often emphasized the manner in which the location of a key industry, which may have been more or less fortuitous, brings about a geographical association or agglomeration of minor or satellite industries. Convenience is an important factor in such a development, but of perhaps greater importance is the increasingly effective cross-fertilization between industries as contributors to the general pool of skills, knowledge, and facilities. The region which enjoys, for whatever reason, the "momentum of an early start" often comes to possess decided and cumulative advantages, by comparison with other regions, which can only be explained by the priority of industrial activity. In the long run, no doubt, the major determinants of industrial location will correct the balance; but, for the time being, the advantages of priority will remain. In relation to the West, California has always enjoyed the momentum of an early start and the first problem, therefore, is to explain how and why it was that western industry struck its first roots in the state.

The isolation of the West does not alone account for California's industrial head start. The Spanish colonists were more dependent on the mother country for manufactured products than were the early settlers on the Eastern seaboard. More than anything else, it was the discovery of gold that got California off to a head start in the competition for industry. The suddenness with which the population increased created an immediate demand for manufactured articles of all kinds and the long, slow, costly, and uncertain trip around Cape Horn or across the Isthmus of Panama precluded the importation of articles and products with sufficient swiftness and volume to meet the demand. Costs were not a factor in the problem. All sorts of articles, badly needed, were simply improvised in California regardless of unfavorable cost factors. Flour mills, canneries, sugar refineries, tanneries, and foundries came into being overnight to meet insistent demands, and once established, they enjoyed for two decades the protection which distance from eastern manufacturing centers provided.

In relation to mining equipment and machinery, in particular, California enjoyed a marked advantage. Prior to 1848 mining had not been a major factor in the American economy, and little was known about the manufacture of mining equipment. In fact, mining machinery was a California invention. The conditions under which minerals were exploited in California, moreover, were novel and therefore challenging to the inventive faculty. New methods and new equipment had to be constantly improvised to meet new conditions. As the mining frontier expanded, the demand multiplied, as Bancroft noted, "for peculiar implements and machinery." Being of great bulk and weight, mining equipment was difficult to transport. The delay in securing new parts for repair or replacement alone precluded the possibility of using eastern products. The increased cost and risk of shipment during the Civil War also stimulated local manufacture. The year following the discovery of gold saw the first foundry established in San Francisco. By 1866, some 13 iron foundries and 30 machine shops in San Francisco, employing a thousand workers, were producing castings of an annual value of two million dollars; and there were, in addition, some 23 iron foundries in other parts of the state.

The manufacture of mining machinery in California was a highly specialized and diversified industry. Deep mining required novel machinery, often specially designed for a particular mine. It brought into being an active demand for hydraulic pumps, air compressors, and hoisting gear. Different kinds of ore had to be treated by different processes and, since the varieties were numerous, the number and type of crushers and amalgamators multiplied. New explosives were constantly introduced, and high-pressure accumulators brought about the use of hydraulic power which, in turn, stimulated the demand for pipe. A by-product of other mining operations, the discovery of lead made possible the establishment of shot towers and lead works. The difficulty of mountain transport and the irregular topography of San Francisco suggested the feasibility of cable-roads, and thereby increased the demand for wire-rope and pulleys. "The California gold mining region," write Messrs. Miller and White, "was the proving ground for mining methods that were later used, with necessary adaptations, in other mineralized regions of the world." By 1870 San Francisco was not only the center of the manufacture of mining equipment in the West, but it had begun to export mining machinery on a large scale. It is important to note that the *diversity* of conditions is what made California the ideal "proving ground" for the manufacture and design of mining machinery.

The development of the tom, rocker, pan, and sluice-box in California stimulated the demand for timber, lumber, and sawmills. The dryness of the climate made the use of lumber feasible for many operations which elsewhere would have required other materials. Lumber was needed for flumes, ditches, sluices, aqueducts, and windmills. The V-flume and the loading chute—two California innovations—created a great demand for lumber and in turn led to the invention of the adjustable saw tooth and the triple circular saw. Mills and shops multiplied as rapidly as sawmills. Many types of novel lumbering equipment, such as gang-sluicing machines, guides, levers, and pulleys were required to handle the enormous redwood and other types of trees. Most of this equipment was, of necessity, manufactured in California. Here the same *diversity* of conditions provided a powerful stimulant to local manufacture and design. Sawmills and shops which had come into being to satisfy the demands of the mining industry were later used to manufacture crates, boxes, frames, and finished timber.

Wagon-making and shipbuilding developed, one might say, as by-products of the initial stimulation which mining provided. Foundries established to manufacture mining equipment could be, and were, quickly converted to the manufacture of other products. In fact, it would be difficult to catalogue the industries which were indirectly stimulated by mining. The list would include transportation, shipbuilding, food processing, lumbering, fishing, and other industries. Furthermore mining made possible the accumulation of capital which, in turn, made possible an expansion of all kinds of services and facilities.

The discovery of gold, in combination with other factors, notably the isolation of the West, gave California a distinct head start over the other western states as a center of manufacturing. Once these nascent industries were established, they had the effect of attracting other industries. It was, above all, the cultural *peculiarities* of the things demanded—novel forms of mining equipment and lumbering equipment—which provided the stimulus for industrial activity. It was for this reason, very largely, that California became a manufacturing center almost at the same time it became a State. The rapid growth in population explains the demand, but it was the novelty of the environment that stimulated local invention and manufacture. The primary dynamic of industry in California might, therefore, be said to lie in the novelty of the environment.

THE EDGE OF NOVELTY

If one jumps from 1848 to 1948, from mining machinery to furniture and sportswear, the same peculiar dynamic may be discovered in the expansion of California industry. Today Los Angeles ranks second only to New York as a garment manufacturing center. Almost two-thirds of the state's garment-making industry is concentrated in Los Angeles where a thousand or so manufacturers, mostly small operators, are engaged in the industry. Between 1936 and 1944, the industry showed a 475 per cent increase in volume of production. Nowadays some 3,000 apparel buyers troop into Los Angeles each year to place orders for spring and summer garments, primarily sportswear, whereas a decade ago most of these buyers would have gone to New York. By 1944 the Los Angeles garment industry employed 35,000 workers, turned out a product worth $265,000,000, and was selling 85 per cent of its products *east* of the Rockies. How is one to account for this phenomenal increase?

The word "sportswear" is the key. The impress of California styling in clothes first became noticeable about thirty years ago, and, primarily in connection with sportswear. Novel conditions of living, reflecting climatic differences, created a compulsion to invent something new and different in the way of clothing. California manufacturers began to meet this need by designing new types of sportswear which, being better adapted to local conditions than the standardized products offered by eastern manufacturers, promptly found a market. Certain of these products gradually began to move eastward, carrying the California label, and, here and there, small shops were opened in eastern cities for the sale of "California Sportswear." In a rather insidious manner, the word "California" became associated in the public mind with the word "sportswear." The success of these new designs in California cannot be fully explained merely by noting that they were better adapted to local conditions. The *willingness* of the Californians to try them was also a factor. California *is* different from Iowa, and this difference means that it is possible to dress differently without being regarded as a "crank" or "freak." This willingness to experiment, to try something new, has also served as a stimulus to the designers.

When California sportswear first began to invade the eastern markets, New York manufacturers made the mistake of assuming that it would not "catch on" as a national fashion fad. This is the mistake that the nation has so consistently made about California, and it is one of California's secret trade weapons. Never make the mistake of assuming that what works in California will not work elsewhere for the exact opposite is nearer the truth. Los Angeles garment manufacturers, being aware of this trade secret, were able to avoid direct competition with eastern manufacturers by concentrating on sportswear. By concentrating on bold, original designs in women's casual clothes, they not only took possession of a largely non-competitive niche in the market, but initiated a trend toward casualness in clothes which is now nation-wide. What started out as a California "fad" has become a nation-wide fashion.

To understand the underlying cultural dynamic, however, a further finesse must be noted. Garment design in New York is not primarily geared to need but is largely determined by more or less accidental appraisals of what is or might become fashionable. In New York, as one manufacturer has noted, a good design may be something "picked out of the air or picked up in Paris." California designs, on the contrary, are based on need. This is not to say that California designers are smarter than New York designers, but it does imply that California has a compulsive environment. "Anything and everything" simply will not work in California; design must be based on function and need. This is what the Los Angeles designers mean when they say that California manufacturers are "less traditional, less conventional," than eastern manufacturers. The fact is that they *had* to be less traditional in order to gain a foothold in the local market; the lack of conventionality is not studied or invented—it is born of necessity. And a nice paradox is involved here. For "what works" in California is likely to succeed elsewhere precisely because it was designed to meet a specific need. Seymour Graff, head of the California Apparel Creators, has summarized the history of garment-making in this statement: "We were originally small manufacturers in an area where styling had to be 'different' if we were to exist."

The same dynamic may be noted in furniture manufacturing. In 1923 forty furniture and bedding manufacturers in Los Angeles did an annual wholesale business of about $15,000,000; today 350 local manufacturers produce products worth approximately $167,-000,000 annually. Here, again, the increase in population locally does not fully account for the success of the industry. The key to the success of furniture manufacturing in Los Angeles is to be found in the obvious fact that an overstuffed divan in Southern California is about as useless as a snow-plow. Californians like modern furniture because it offers a better solution to the problems of living than the standardized products of Grand Rapids. Both the need for, and the willingness to accept new furniture designs have been a primary factor in the development of the local industry. In 1948, 60 per cent of the west coast furniture stores reported modern furniture in top demand by comparison with 39.3 per cent of the furniture stores in New England. Not all of the furniture manufactured in Los Angeles is "modern" but the bulk of the product is better adapted to living conditions in California than the eastern importations. Furniture designers report that the California designers are "bolder in their actual use of color" in upholstered furniture, drapes, lamp shades and accessories than designers in other areas. The climate is doubtless a factor in accounting for this difference but people somehow "feel" different in California and want to express this feeling in their homes. Whatever the cause, it was by developing products aimed at meeting specific local needs that the furniture business got started in Los Angeles.

Novelty of the environment appears as a dynamic in many different fields of industry in California. For example, western truck operators have long bemoaned the fact that trucks are designed and built in eastern states. These same operators contend that California trucking problems are unique. Extremes of altitude and temperature, and other factors, create many special problems. For years the fleet operators sought a meeting with the design engineers on the assumption that trucks designed to meet California conditions would function with great all-round efficiency in other areas. Such a meeting, with the Society of Automotive Engineers was held in San Francisco in 1948.[1] It is the versatility of California's environment that makes it the ideal "testing ground" for all kinds of industrial designs and for design in general. But, as Gregory Ain, the Los Angeles architect, has pointed out, climatic differences alone do not account for the Californians' preference for modern design. "Eastern traditionalists," as he has said, are less willing to experiment than California architects, just as California home-owners are more willing to try "something different" than home-owners in other areas.

The fact is that Californians have become so used to the idea of experimentation—they have had to experiment so often—that they are psychologically prepared to try anything. Experience has

[1] Los Angeles *Daily News*, July 30, 1948.

taught them that almost "anything" might work in California; you never know. "Whenever we have had anything new to try out," reports A. O. Buckingham, chairman of Brand Names Foundation and vice-president of Cluett, Peabody & Company, "I have always asked our people to send it out here (Los Angeles) because I knew you would try it. I knew that you would not think of all the reasons that old established communities can think of why it wouldn't be successful, but that you would take it and say: 'Let's try it.' After you have tried it and made a success of it then the old established cities accept it. If anything, that is the magic which has made Los Angeles great." [2] There is, of course, no single explanation for this widespread experimental attitude. It reflects the newness, the diversity, and the representative character of the population but it also reflects the environment itself. In explaining why the Carnation Company had established headquarters in Los Angeles, Paul H. Willis, its general advertising manager, said: "My best theory is that here in Los Angeles there is a fusion of regions and cities. Nowhere else have I ever met so many people from other places. My point is, that such a fusion of ideas, cultures and skills is productive of activity, growth and progress. It is what makes Los Angeles a dynamic city." [3]

One does not exhaust the theme of California's novel environment, physical and social, as a factor in industrial growth by simply noting its effect on design and markets; the implications are, in fact, almost endless. For example, it has been noted that the industrial growth of Los Angeles has not been accompanied by the same degree of specialization that normally goes with economic maturity. Dr. Philip Neff has worked out a scale for measuring economic specialization on a comparative basis. According to this scale, Los Angeles in 1940 had a "specialization score" of 23 percentage points by comparison with Pittsburgh's 36 and Chicago's 35. One of the reasons for the diversity of manufacturing activities in Los Angeles is that the adaptations to meet *peculiar* local needs are more readily made by the local manufacturer since his entire output can incorporate the desired changes; whereas the

[2] *Ibid.*, September 11, 1946.
[3] Los Angeles *Times*, September 8, 1948.

large eastern manufacturers hesitate to modify a product which has been standardized to meet large-scale production requirements. In more than one case, the eastern product has been standardized to meet conditions which do not prevail in the Southwest. This is what the more perceptive eastern manufacturers, who have been locating plants in Los Angeles, have in mind when they say that "the Pacific Coast is a separate empire, in more ways than one."

The industrial expansion of Los Angeles in the last decade has been characterized: first, by its wide diversity; and, second, by the number of small plants. Of some 95,770 businesses in Los Angeles County, 92,000 employ fewer than 50 workers and 79,838, or 83.4 per cent, have less than 10 employees. The relatively large number of small concerns is, of course, an attraction to the manufacturer. When numerous peculiar local needs are to be met, new opportunities are created for concerns which could not hope to compete with mass-production firms in the East and Middle West. Many of these concerns remain small in size because they produce for the local market. In short, it is the cultural situation which creates special opportunities for the small plant catering to peculiar local needs, and it is the peculiarity of these needs which throws up a kind of invisible "protective tariff."

LOCAL NEEDS AND NEW MARKETS

Since California's environment is novel, it has always had a disproportionately large share of what are known as resource-based activities; that is, activities based on peculiar local needs. The extraordinary amount of electrical power used in California, both in industry and agriculture, has created a large and vigorous pump and pumping-equipment industry which is highly localized. The same is true of the manufacture of many electrical appliances. Localization of the canning industry has brought into being an industry which manufactures a wide variety of novel canning appliances: peach slicers, closing machines, cherry-pitting equipment, conveyors, fruit-marking machines, raisin steamers, peach defuzzers, milk equipment, syrupers, and bottle fillers. Many of these appliances are not only specialized, but are made to order and are used only in California. The industry, it has been said, has developed primarily in response to "the peculiar needs of the food processors of the state." The manufacture of oil field machinery and tools, one of California's strongest industries, is highly localized. Although the manufacture of mining machinery has declined in importance, it, too, is a localized industry chiefly concerned nowadays with the manufacture of special-purpose products. Since the mildness of the climate favored the development of water-heating units not associated with central heating plants, an important industry has grown up around the manufacture of oil-and-gas-fired water heaters. Similarly the manufacture of evaporative coolers, well-adapted to the warmer and drier areas of the Southwest, has become an important local industry.

As a result of the peculiar needs of California agriculture, many new types of farm machinery and equipment have been developed in the state. Stockton, California, was one of the first centers for the manufacture of farm machinery and equipment in the United States. Among the products so developed were: heavy-duty offset discs, deep tillers, subsoilers, special type potato diggers, bed shapers and irrigation equipment, green-crop loaders, hay loaders, orchard brush clippers, bale pickups, weed burners, tree shakers, field wagons and loaders, and smudge pots. As many as 60 or 70 custom-made combine harvesters were produced annually in Stockton in the 1880's. One of the larger local concerns of this period originated in the barn of a farmer who made a heavier-than-standard disc for his own use; requests from his neighbors for similar equipment led eventually to the production of more than 1,000 discs a year. In fact the reason that California did not develop into a large farm-equipment manufacturing center was that so many of its products were specialized-in-use. But the manufacture of equipment, in large part custom-made for peculiar local needs, is still an important industry. The following is a list, selected to illustrate the diversity-in-need, of special farm equipment developed and manufactured in California: bale loaders, sack loaders, sugar beet planters, tree shakers, elevators, sprayers, bean cleaners, tray lifters, pig brooders, fruit dehydrators, shaking towers, rice driers, electric poultry brooders, walk-in refrigerators, milk coolers, quick freezers, flax harvesters, bulk-handling equipment, buck rakes, special crop harvesters, desert coolers, nut dehydrators, depth controls for disc harrows, flexible harrows, orchard heaters, barn equipment, and dobeakers.

The comparative uniformity and mildness of the California climate as well as the great variety of scenery were factors in the location of the highly important motion picture industry which, in turn, has attracted many other industries. In this case, the industry also happens to be well adapted, one might say "especially" adapted, to the Los Angeles area. For, as Dr. Clifford M. Zierer has pointed out, the *peripheral* position of Los Angeles in relation to the major national market was not a handicap in the manufacture of motion pictures. Film prints move to market with great rapidity and little cost, and few raw materials are required in their manufacture. A physical and social environment favorable for attractive living was still another factor in the location of the industry. "If the industry had not gone through its 'outdoor' and 'western' stages of development," reports Dr. Zierer, "it is more than likely that it would never have left the eastern half of the country." One could trace a similar cause-and-effect pattern in the location of the radio industry. The radio and motion picture industries, in turn, have been attracting related industries, such as the recording industry and the manufacture of band equipment. The manufacture of airplanes is still another resource-based industry; in this case the resource is climate. Sales of four major Los Angeles aircraft plants in 1948 totaled more than $300,000,000. Currently 55 per cent of the government's orders for new planes are filled in Los Angeles, by comparison with 26 per cent in wartime. The airplane industry, in turn, has been a factor in the development of a western steel industry and in the retention of the aluminum industry since the end of the war.

Peculiarities of the west coast market are also a factor in the location of industry. Since west coast families are likely to be smaller than the national average, they require more household gadgets and home furnishings. A million mid-western residents will not provide a market for as many refrigerators, washing machines, and electric irons as a million west coast residents. Furthermore, when people move from one area to another they generally sell their old belongings and buy new ones when they reach their ultimate destination. "Whether these new residents want to or not," to quote Dr. David E. Faville of the Stanford Graduate School, "they usually spend as much in the first six months of getting settled as the old residents spend in six years." Some 32,000 net additional trade outlets were established in California during the first ten months following V-J Day. As many as 5,000 new outlets were opened in a single month.

Van Beuren Stanbery, special Department of Commerce representative on the west coast, has pointed out nine exceptional characteristics of the California market area: a high ratio of employment to population brought about by reason of the concentration of population in the active working ages, between 20 and 45; high *annual* wage incomes due, first, to higher hourly rates and, second, to the fact that in many outdoor occupations, where weather is a factor, men can work many more days per year; the existence of important extractive industries, such as agriculture, mining, forestry, and fishing, which produce high financial returns in proportion to the number of people employed; a highly urbanized population (urban incomes are generally higher than rural); the momentum of an early start in industry which has made Los Angeles and San Francisco important centers of finance, distribution, and services for the entire West; a high concentration in distribution and service activities which produce high per capita incomes (64 per cent of all employment in California is in these categories: the highest proportion for any state); a mounting tourist trade which helps to support employment of trade and service workers; the existence of industries, like the motion picture, canning, and petroleum industries, which bring in large revenues from beyond the borders of the state; and the presence of many wealthy retired people who draw income from outside the state. For these and other reasons, California experienced the greatest rate of income growth in the period from 1929 to 1945 of any large area in the country. The average per capita income payment in the nation in 1940 was $579; the average for California was $811. For many types of services and home-market (that is, residentiary) industries, the presence of people is per se the decisive locational factor. Not only has California continued to grow at a phenomenal rate, but, over the years since 1910, the percentage increase in the number of households has been more rapid than the rate of population growth.

The *peripheral* position of California, which would normally be a decided disadvantage, is actually a marked advantage when considered in light of the remarkable population growth of the state and the exceptional qualities of the California market. For the same factor of distance, which removes California from the eastern centers of population, also handicaps eastern businesses and industries that want to tap the highly favorable California market. High freight rates, which are a disadvantage in one sense, are, in another sense, a favorable factor. As transportation costs increase, more and more manufacturers are finding it advantageous to establish local plants in California rather than to produce goods elsewhere for shipment to California. In locating a branch plant in California, officials of the Bristol-Myers Company recently said that they had done so "because the Pacific Coast is a self-contained unit—wholesalers don't go east of the mountains—and because it represents a range of climatic conditions and living habits."

BULL OF THE WEST

By comparison with the other western states, California has always possessed exceptional financial power which has been a factor in the momentum of its early start in industrial activity. The banking and financial history of the state, as one might expect, is highly exceptional. From the earliest date, the Californians have been "hard money" advocates. The first constitution prohibited the issuance of state bank notes and, during the Civil War, California refused to accept the depreciated "greenback" currency of the federal government. Down to 1918, unusually large amounts of gold and silver circulated in the state; "pennies" were referred to as "junk" and, when offered in change, were often left lying on the counters.

Historically the basis for this persistent preference for "hard money" is to be found in the circumstance that for several decades after 1848 California had to improvise its own currency. In the first decade, gold dust was the almost universal medium of exchange, with every store and shop having scales in which the miner's "pinch" of gold dust was weighed in the purchase of supplies. The isolation of the state and the delay in making payments for imports and exports required that some type of currency should be improvised and "gold" was it. The use of gold, however, was supplemented by the use of a bewildering variety of foreign coins: rupees, pistareens, Mormon coins, Mexican double-reales, and coins minted by private concerns. However, the trafficking in gold dominated all channels of finance. Not only was gold the most commonly used medium of exchange but the business of buying and selling gold became quite extensive since a federal mint was not established until 1854. As a consequence "private banks," usually one-room institutions with a counter, safe, and scales, sprang up overnight. Banks of this sort emerged in San Francisco, as Benjamin C. Wright pointed out, with "a suddenness that would have been considered startling under any other circumstances in a city of the same population." The famous early-day express companies, such as Wells, Fargo & Company, also performed a banking function. During the Civil War, when commerce with the rest of the country was sharply curtailed, California was the only state which remained on a gold basis. The fact that it remained on a gold basis served to attract large sums for investment from foreign countries and, for several decades, foreign bankers, notably British bankers, were an important element in financial circles. Later, with the discovery and development of the Comstock Lode, large sums derived from mining were invested in banks and lending institutions such as the Nevada Bank, formed in 1875, which was owned by James C. Flood, W. S. O'Brien, John W. Mackay, and James G. Fair, the famous Comstock Lode tycoons.

Having improvised their own currency, the Californians were extremely reluctant to accept any form of paper currency. The "hard money" policy of the state had protected the economy against the ravages of wild-cat bank notes issued by state banks; and, also, had safeguarded the economy against the depreciated greenback currency of the Civil War period. The Californians, also, were extremely skeptical of banking corporations and, in fact, of all corporations. This skepticism found expression in a provision in the first state constitution establishing the principle of the individual liability of stockholders in corporations, the first provision of its kind in the United States. For many years the people would not permit the incorporation of banks but, once incorporation was permitted, the state was extremely lax in the supervision of state banks. As might be expected, however, once the necessity for regulation was demonstrated, the state in one bold act brought itself not only abreast of other states in the regulation of banks but, in the adoption of a comprehensive banking act in 1909, pointed the way to the other states.

As a result of this "peculiar" financial history, based on the discovery of gold, capital has shown a consistently rapid rate of accumulation in California. High per capita income, the urbanization of population, the migration of people with means, and other factors have contributed to the rate of capital accumulation. In 1925, California ranked eighth among the states in per capita savings; tenth in number of savings depositors; and fourth in the gain in savings on a per capita basis. Mining fortunes and foreign investments, represented by the appearance in early San Francisco of banking firms with such names as the Commercial Bank of India, the British & California Banking Company, the Anglo-California, and the London, Paris & San Francisco Bank, and many others, made possible a fund for investment which had a stimulating effect on local industry. In 1865 San Francisco ranked ninth among American cities in the amount of capital invested in manufacturing ($2,211,300) and in the value of manufactured products ($19,-318.74) which, as Dr. Cross has pointed out, was certainly a remarkable showing for "such a young city in a frontier community." In large part, the funds for this nascent industrial development were made available by California investors and lending agencies.

A "colony" California may be, in certain respects, but it is indeed a strange colony which can boast that it is the domicile of the world's largest bank, the Bank of America, which had assets in 1948 of $5,859,234,000. Today this one bank has over 500 branches in 162 communities in the state, 3,500,000 depositors, 13,000 employees, and holds 40 per cent of the state's bank deposits. Both the Bank of America and Transamerica are western-controlled; they are, in fact, controlled by the Giannini family. Nowadays, also, Transamerica controls the First National Bank of Portland with 28 branches; a chain of banks in Nevada with 75 per cent of the state's deposits; and rapidly expanding banking systems in Arizona and Washington. This is quite an institution to emerge in the western "colonial empire"; but just what has been the secret of its phenomenal rise to power?

Amadeo P. Giannini, "The Bull of the West" in Matthew Josephson's phrase, was born in California in 1870, the son of Italian immigrant parents. By the time he had reached thirty years of age, he had made a sizable fortune in the produce business, the economics of which he had thoroughly mastered. On October 17, 1904, he opened the Italian Bank of California, a one-room bank with a capitalization of $150,000 which catered to the needs of the "North Beach" Italian colony of San Francisco. It should be emphasized that Giannini's training was in the produce business; prior to 1904, he had had no experience or training in banking or finance. As a matter of fact it was precisely because he was an "outsider" to the banking world, completely unhampered by the "traditions" of banking, that he was able to see what the traditional bankers could not see, namely, the opportunity for branch banking in the self-contained economic empire that is California. Although branch banking did not originate in California, Giannini was the first successful branch banker in America and, today, the Bank of America is the only state-wide branch banking institution in the United States.

The opportunity for branch banking in California existed by reason of two facts. In the first place, California, unlike many of the other states, had never specifically forbidden branch banking; in fact state regulation of banking hardly existed prior to the adoption of the banking act of 1909. This act, which obliterated all prior legislation on banking, specifically authorized state-wide branch banking and, with the adoption of this act, Giannini began his amazing march to power. In the second place, the economy of the state has always been ideally adapted to branch banking. California is a self-contained empire with a highly diversified and intimately inter-related economy. If the financial power of the state could be integrated on a state-wide basis it would be possible, so Giannini reasoned, to carry the grower in Sacramento whose peach crop had failed by offsetting this loss against, say, the bumper potato crop in Kern County. It will be recalled that the agriculture of the state is not only highly diversified but that it is highly specialized by area. With most of the farmers in a particular area specializing in the production of a single crop, a great peril faced the local banker. The only way by which the risks inherent in this type of economy can be equalized is on a state-wide basis.

Not only was the small rural banker faced with the problem of securing a diversification of risks, but the highly seasonal character of agricultural production created a situation in which there was little demand for loans at certain periods of the year and such a demand for loans "during the season" that the local banker had to seek outside assistance. By synchronizing seasonal demands on a statewide basis, Giannini was able to shift funds from one part of the state to another, thereby making possible the maximum utilization of these funds. There was, also, a marked discrepancy in the availability of investment funds and the need for these funds in California. The older, settled communities had a surplus for investment; the newer communities a deficit of funds and an enormous demand. Statewide banking made possible a leveling off of these discrepancies. Since an exceptionally large proportion of California's agricultural production has always been exported, and consists in cash crops, the financing of this production has presented peculiar problems and, conversely, has created special opportunities. The rapid rise of large cities, like Los Angeles, spread over a large area, also created an opportunity for branch banking since, under this system, it is possible to extend branches into communities in which the volume of business would be too small to warrant a unit bank. Finally the migration of people of means to California has involved a migration of capital and savings which, in many cases, can be most effectively gathered by branch banks widely dispersed on convenient street corners. This was the opportunity that Giannini saw and the fact that he transformed the Italian Bank of California, known as the "Baby Bank" of San Francisco, into the largest bank in the world in the period from 1904 to 1940 is the best evidence of how thoroughly he had grasped the situation. In a sense, therefore, branch banking is a California innovation.

Today, as yesterday, California continues to attract out-of-state capital like a magnet. Vast sums have been poured into the state in the last few years for all types of investment.[4] The Prudential Insurance Company, which now has headquarters in Los Angeles serving the eleven western states, has been investing funds in California at the rate of $15,000,000 a month. Providing a primary market for the securities of local corporations, the Los Angeles Stock Exchange in the last year has shown a 40 per cent increase in the dollar volume of business and is rapidly becoming one of the major regional exchanges in the nation. In terms of the dollar volume of general business transactions, Los Angeles now ranks third among the cities. It is today one of the most important insurance centers in America with more than $5,000,000,000 in life insurance in force through ten locally domiciled companies.[5] The rise of western-rooted financial power in California has had, of course, a marked effect on the growth of industry within the state.

[4] "Upsurge of Los Angeles Dazzles N. Y. Realtors" by Stephen G. Thompson, New York *Herald Tribune*, Nov. 23, 1947.
[5] Los Angeles *Daily News*, July 7, 1948, p. 46.

13

THE FABULOUS BOOM

◇◇◇

The West Coast, long regarded by the rest of America as a kind of colonial outpost, can now claim to be a thriving heartland. The three states on the western rim of the nation comprise a new citadel of power.
—KIPLINGER MAGAZINE, February 1948.

CALIFORNIA HAS had many booms in the first century of its existence but none like "the fabulous boom" of the last eight years. For this has been California's first real "industrial" boom; its first real challenge to eastern industrial supremacy. During this eight-year period, the federal government has spent more than a billion dollars on the construction of new industrial plants in California, and $400,000,000 have been invested for the same purpose by private capital. Los Angeles, which had never been thought of as an industrial center, found itself handling more than ten billion dollars in war production contracts. With a population increase of between 30 and 35 per cent, total non-agricultural employment in the three west coast states gained 42 per cent for the period, and manufacturing employment increased by 66 per cent. More astonishing than this spectacular upsurge in industrial power was the fact that, contrary to all expectations, the severe cutbacks in employment which were anticipated in the period from 1945 to 1948, failed to take place. On the contrary, the trend during these years was consistently upward. The peak of the expansion was not reached, in fact, until the end of 1948.

The wartime expansion of industry on the west coast has a special significance that must be understood before one can properly appraise the present status of western industry. Unlike other areas, the West did not *convert* to war production for there was nothing much to "convert"; what happened was that *new* industries and *new* plants were built overnight. It is most significant, therefore, that the expansion in industry was in new lines of industrial production, notably in the production of durable goods, and not in typically western industries, such as food, forest, and mineral products. Manufacturing employment rose in California by 75 per cent so that, by 1948, 16 per cent of its non-agricultural employment in manufacturing is in industries other than food and forest products. As a matter of fact, only 20 per cent of California's permanent manufacturing expansion took place in farm and forest products, which accounted, however, for 63 per cent of Oregon's and 30 per cent of Washington's expansion. What this indicates, of course, is that a much larger proportion of wartime growth in California was in durable goods manufacture than was true for the nation. Still lagging behind areas of heavy industrial concentration, California made seven-league strides during the war on the road to industrial maturity. It is equally important to note, however, that the wartime expansion of industry on the west coast merely accelerated a long-term trend; war or no war, the expansion would have occurred but, as a result of the war, the process was greatly foreshortened. A brief backward glance at the development of industry in California will serve to underscore these conclusions.

THE LONG-TERM TREND

Despite early industrial activity in San Francisco, industry had little general significance in California until the turn of the century. From 1900 to 1940, however, a remarkable transformation took place in the state's economy. In this forty-year period, the state changed from a raw-material economy, with emphasis upon agriculture, mining, and lumbering, to an economy in which the value added by manufacturing accounted for two-thirds of the entire value of basic production. Most of this development, moreover, took place subsequent to 1910. Prior to 1910, the absence of an inexpensive source of power had been the main limitation on industrial expansion. But by 1910 the development of oil and hydroelectric power had reached a point which enabled the state to offset somewhat higher costs of production. The first hydroelectric plant in California was installed in San Antonio Canyon, in Southern California in 1891; by 1920, California held the first rank among the states in the development and utilization of electric power. Once oil and electric power provided the missing energy base, California began to experience a minor industrial boom.

The pre-war industrial trend can best be illustrated by developments in Los Angeles, long regarded as the industrial pygmy in the roster of California cities. Despite this general impression, however, manufacturing had shown a more remarkable development in Los Angeles in the period from 1900 to 1940 than in any other urban area in the nation. Between 1899 and 1914, the number of manufacturing wage earners in Los Angeles increased at an annual rate of 11 per cent, by comparison with a 7.5 per cent increase in Detroit. In 1920, Los Angeles had 17.5 per cent of the state's total employment; in 1940, 41.1 per cent. At the turn of the century, Los Angeles had only 6,600 manufacturing wage earners out of a population of 170,000; but by the end of 1947 this figure had risen to 241,000, almost double the figure for the San Francisco Bay Area and representing nearly one-half of the state's pay roll. During the period from 1900 to 1940, the spectacular increase in population consistently obscured the remarkable increase in manufacturing. Since the city continued to have a smaller proportion of its total population engaged in manufacturing than other cities of comparable size, the impression prevailed that Los Angeles lacked a significant industrial base. Actually the expansion in manufacturing was as remarkable, in a way, as the increase in population.

In the post-war years the population of Southern California has been increasing at the rate of 250,000 annually; and it has been this increase which, to a large extent, has prevented the cutbacks which everyone anticipated. As a matter of fact, industrial growth proceeded in the period from the end of the war up to March 1948, at a rate 23 per cent greater than the wartime rate of growth! By August 1948 general employment had reached a figure higher than the peak of wartime employment, and per capita industrial investment for the post-war years averaged $14—nearly three times the figure for New York City.[1] The continued expansion of industry in California since the war certainly challenges the thesis that it was the wartime boom which alone changed the character of the economy of the state.

In considering the long-term trend of industry in California, it is important to note that Los Angeles has rapidly outdistanced the San Francisco Bay Area. Among the major industrial groups recognized by the Bureau of the Census, the Bay Area tops Los Angeles in only one category at the present time. Originally San Francisco had "the momentum of an early start"; now this momentum has shifted to Los Angeles. Nowadays San Francisco suffers from its cramped peninsular site and inferior rail connections.[2] The spread-out, highly dispersed geographical pattern of Los Angeles manufacturing happens to square with the modern trend toward uncrowded, one-story manufacturing plants located on the periphery of built-up areas. Not only does Los Angeles have better transportation facilities than San Francisco, and a larger market area, but it has space to burn. Climate has also been an important factor. The important aircraft industry, for example, completely bypassed the fog-ridden Bay Area. But, more important than the relative absence of fog, has been the existence of vast empty spaces of the nearby Mojave Desert which, as major testing and experimental laboratories, have been of great importance to the aircraft industry. Southern California is today the nation's leading center of aerodynamic research. The aircraft industry, also, happens to be one which has exceptional space requirements and there is plenty of "space" in Los Angeles. The motion picture industry is another industry in which the climate-and-space factor is important, and with motion pictures have come many subsidiary industries, such as, the manufacture of photographic equipment.

Paralleling the growth of the aircraft industry has been the emergence of Los Angeles as an important automobile center, second only to Detroit. In 1941 four Los Angeles assembly plants produced 154,000 cars; in 1948 the capacity was more than 650,000. With the assembly plants have come plants manufacturing

[1] *New York Times*, March 29, 1948.
[2] See: *Geographical Review*, Vol. XXXIX, pp. 229–241.

parts and accessories. By the end of 1948, Ford's west coast purchasing program was operating at the rate of 45 million dollars a year, with nearly 50 independent manufacturers participating. Ford, which now assembles more than 14 per cent of its output in California, estimates that transportation of parts to the west coast adds an additional 10 per cent to its costs and involves a ten-day inventory delay in shipment west. In these and other developments, Los Angeles has had many advantages to offer: lower building costs (an aspect of the spread-out character of the city), excellent rail and trucking facilities, and lower costs occasioned by the fact that many activities can be carried on out-of-doors and with a greatly reduced requirement for heating in industrial plants.

Behind the "arrival" of Los Angeles as a manufacturing and industrial center is a factor which is often overlooked, namely, that the rapid development of industrial technology makes for an equally rapid rate of obsolescence. As Thorstein Veblen pointed out, Germany had a marked advantage over England once it started to build an industry since it could apply the latest technological processes in a more thorough-going manner. This was the penalty, as he put it, which the British had to pay for taking the lead in the development of industry. There was in Great Britain, in other words, "a fatal reluctance or inability to overcome this all-pervading depreciation by obsolescence." What Veblen had to say of Germany's advantages in relation to Great Britain can be said of the West's advantages in relation to the East, or of Los Angeles advantages in relation to San Francisco. Los Angeles just happens to be the first modern industrial city in which industry is widely dispersed. Because of its late start, by comparison with older industrial areas, it has been able to build *modern* plants and to profit by the experience of industry in older centers. The new scientific technology has a particular relevance to the West since so many of the West's resources involve the use of new metals, new chemicals, new processes. Technologically, the East's headstart in industry could prove to be the West's major advantage.

In addition to branch plants, local assembly units, and retail outlets, many eastern and middlewestern concerns have opened "regional" headquarters in Los Angeles. Here, too, one can note the "pull" of cultural factors. In selecting regional headquarters business concerns are primarily concerned, of course, with basic factors, such as transportation facilities, proximity to the trade, and so forth; but they are also influenced by secondary factors, such as, the pleasantness of the environment. In explaining why Prudential Insurance Company had established regional headquarters in Los Angeles—a move that brought 1,116 Prudential employees to the city—the president of the company stressed the fact that the company had insurance in force on the lives of 2,500,000 people in the eleven western states and Hawaii. Obviously it was logical, therefore, to open a regional office in the West; but why Los Angeles which is less centrally located than Salt Lake City? The answer, I suspect, is to be found in the fact that Prudential executives would much rather live in Los Angeles than in Salt Lake. The same is true, also, of many federal agencies which have located "headquarters" in California. The development of Los Angeles as a "regional headquarters area" has been and will continue to be an important factor in the city's growth. California now leads all the other states in the number of federal civilian employees—202,572 in September, 1949.

Just as the "spread-out" character of Los Angeles has lent itself to a new type of industrial development, fitting in with modern industrial methods, so it has also been an important factor in the development of new marketing techniques. Not only has Los Angeles grown rapidly in the last eight years, but it has grown horizontally rather than vertically. The spreading-out of population has created an enormous opportunity for the establishment of new retail outlets rather than the expansion of existing facilities which would have been the case had the new population been crowded into already established residential areas. There are 154,690 retail outlets in Southern California—within 16,000 of the total number of outlets registered for the entire state in 1943. Paralleling the horizontal expansion of the city has been a development of new marketing techniques. Super-markets were a flourishing industry in Los Angeles in the 1920's before other areas had even begun to turn to this new form of retail selling. Super markets, of course, are a product of the year-round mildness of the climate, high per capita automobile ownership, and the rapid development of out-lying residential shopping districts. Of 1,666 super-markets in the eleven western states in 1939, 1,423 were located in California. The super-markets, it should be noted, lead all other retail outlets in the variety of products handled.

If one places "the fabulous boom" of the last eight years in the perspective of the long-term industrial expansion of California, and also relates it to the peculiar cultural factors which exist in the state, then it is apparent that the industrial upsurge of California is not a transient phenomenon but an aspect of "normal growth." With the approaching industrial maturity of California, the bal-ance of industrial power in the nation has been disturbed if it has not been changed. The cultural factors underlying California's in-dustrial expansion are constants; they will be as influential twenty years from today as they have been in the past. It should be noted, also, that the most important advantage which California possesses as a potential industrial center—namely, its relation to the Far East—has not yet become of major importance. In an industrial survey of San Francisco, Lawrence E. Davies has pointed out that the west coast industrialists are banking for the immediate future, not on the hundreds of millions of potential customers in the Far East, but on the West's population increase from 14,000,000 to 18,000,000 in the last eight years.[3] No one can appraise the Far Eastern potential but it looms large on the horizon. Similarly such developments as the recent completion of a new rail link between Baja California and the Mexican mainland, and the proposal to construct special "trailer body carrying ships" which would make it possible to ship merchandise across the Tehuantepec Route in Mexico, with a saving of 1,500 miles on the present sea routes be-tween the Atlantic and the Pacific, have a great potential in terms of trade with Mexico and Central America.

THE EMERGENCE OF WESTERN STEEL

Of particular importance to California industry has been the construction of the west coast's first integrated blast furnace and

[3] *New York Times*, March 2, 1947.

steel mill at Fontana, some 50 miles east of Los Angeles. Most of the coal for this new plant comes from Utah deposits 500 miles distant; the iron ore (hematite and magnetite) comes from the Eagle Mountain open-pit mine on the desert, 153 miles to the east. The ore reserves, which have been proven, are estimated to be sufficient to sustain operations for a period of 50 years at the present rate of consumption. The Fontana plant, and the new Geneva, Utah, plant, together give the West 2.3 per cent of the nation's pig iron capacity as of 1945. For thirty years prior to the war, the West had been trying to get a steel industry; but, overnight, the steel ingot capacity of the West jumped from around 1,000,000 tons to 3,500,000 tons with the construction of the Geneva and Fontana plants. The future of these plants, however, is uncertain. Westerners were dismayed when the United States Steel Corporation purchased the Geneva plant and then proceeded, through its western subsidiary, Columbia Steel Corporation, to acquire control of the Consolidated Steel Corporation of Los Angeles. The acquisition of the Geneva plant by U. S. Steel, the uncertainty occasioned by recent decisions of the Supreme Court on the basing-point system, and other factors, make it impossible to forecast the future of western steel. But the announcement by the Kaiser interests that a second blast furnace will be constructed at Fontana would indicate that the Fontana operation has proven to be financially successful. Prior to the war, the West did not have a modern steel industry; today it has such an industry. If this industry can be expanded, on a competitive basis, California will be well on the road to real industrial maturity; but the "if," in this case, raises a large and currently unanswerable question.

THE QUESTION OF POWER

On March 9, 1948, California newspapers carried headlines about the "power famine" in the state: daylight-saving time had been invoked as an emergency measure, a power "czar" had been appointed, and a rationing of power had been decreed. These emergency measures, moreover, remained in effect until January 1, 1949. The occurrence of this "crisis" came as a great shock to residents of California where cheap power and fuel had long been reckoned as among the permanent assets of the western economy. What had happened to the power, the energy base? The answer is found in the Federal Reserve Bank Report dated December 15, 1948, indicating that although the output of western electrical power had doubled between 1939 and 1944, power demands had increased by 102 per cent, not only wiping out the margin of reserve, but requiring some rationing. Today, with the exception of the water problem, the major limitation on the industrial expansion of California is to be found in the matter of power.

For twenty-five years prior to 1949, California was entirely self-sufficient in natural gas and petroleum. No natural gas had ever been imported; likewise California had supplied all its fuel oil requirements and had, in addition, exported oil. But, recently, Southern California utilities were compelled to build a 1200-mile pipeline to Texas in order to transport 300 million cubic feet of gas daily into the Los Angeles area: five times the amount of energy now generated every day at Boulder Dam. By January 1, 1949, California oil production reached its estimated peak and the experts forecast, from that date, a steadily declining production. Fuel oil has more than doubled in price in the last three years as California has ceased to be an export state, and has begun to import, with ever higher fuel oil costs in sight. Despite extensive exploration during the war, there have been no major oil or gas discoveries in California since 1939. During the war, utility systems in California were compelled to build steam-generating plants with a capacity of 1,750,000 kilowatts; this expansion of steam-generating plants has placed a still heavier burden on fuel oil. By 1954 both the oil and gas production of the state, it is estimated, will be considerably below present levels although the demand will have greatly expanded. Eventually California may have to use coal, imported from Utah, to sustain the output of its stand-by steam-generating plants. Hence the interest of California oil companies in the shale-oil of Utah and Colorado, and in the oil of the Near East. Secondary oil reserves are still great in California—perhaps 27 billion barrels—but, like all oil economies, the end is in sight. A second pipeline is being built to San Francisco.

California has long been, of course, the leading hydroelectric power producing state with steam electric power being used merely to supplement hydroelectric generation. At the present time, the developed hydroelectric capacity of the state is approximately 1,988,095 kilowatts with an undeveloped hydroelectric capacity of 7,100,000 kilowatts. It will take many years, however, for this undeveloped capacity to be made available. Boulder Dam power, available since 1936, has provided something like 24 billion kilowatt-hours of power or the equivalent of 68,000,000 barrels of fuel oil. The firm power from the project is expected to average about 4,115,000 kilowatt-hours per year. At the rate power is being used in Southern California, however, the demand for power is doubling every ten years. Additions to present power plants on the Colorado and projected new developments would greatly increase the available power but, even so, the added energy represents less than 15 years growth if present demands are projected into the future.

On the question of power, Southern California is particularly vulnerable. Load requirements on existing and proposed power projects on the Colorado River south of Lee's Ferry—the projects within economically feasible transmission distance—will, it is estimated, have exhausted by 1960 the entire potential of the river as far as Southern California is concerned. Most of the other sites in the state are too remote from Southern California to be of much value and are, furthermore, embraced within other power markets. The power situation is of particular importance as the one presently available means of reducing the demand for oil and protecting the existing oil reserves. It takes a barrel of oil to produce 500 kilowatts of electrical energy. Most of the studies of the problem which have been made indicate that Southern California will face a critical shortage of power within a decade.

Already California has begun to forage for power and the prospects are not too bright. There are some coal reserves in the state which can be developed, although the coal is of an inferior variety. The nearest significant coal deposits are in Utah, some 400 miles away. Despite the difficulty of importing coal, however, plans are under way to increase the supply of fuel-based energy, and to develop stand-by steam plants which will, in turn, greatly increase the demand for fuel. It has also been suggested that the power system of California might be integrated with that of the Pacific Northwest for the peak-load periods are, to some extent, different. But the Pacific Northwest is now suffering from an acute power shortage and it is by no means certain that substantial sources of additional power can be made available to California. Second only to the water problem, therefore, power is the real problem of California.

THE PRICE OF EXPANSION

Although the industrial expansion of California in the last eight years has been phenomenal, there are certain aspects of this expansion that deeply trouble the more thoughtful Californians. For one thing, manufacturing has been increasing in the state at a consistently slower pace than population growth. In the period from 1929 to 1947, factory jobs increased 66 per cent in California; but the state's population increase was 74 per cent for the same period. In 1947 approximately 6.1 per cent of the civilian population of the San Francisco Bay Area was engaged in manufacturing; the comparable figure for Los Angeles was 6.2 per cent. Although this represents an increase for both areas over 1939, the 1947 Bay Area figure is still well below the 7.1 per cent employed in manufacturing in 1929. In other words, the number of manufacturing employees has increased but not as a percentage of the total population actually employed. For this same period, there was a state-wide increase in factory jobs of 71 per cent, and a 74 per cent gain in population. Generally speaking, wage-earner densities of population are still less in San Francisco and Los Angeles than in comparable areas of the East. Since the primary dynamic of industrial expansion in California has been population increase, it is perhaps inevitable that population should consistently outstrip industrial employment. This discrepancy is already a matter of considerable concern in California. In the event of a nation-wide depression, California would be hard hit and the impact would be immediate.

Of still greater concern than the lag between industrial employment and population increase is the fact that so much of the industrial expansion in California of the last eight years represents the establishment of branch plants. Between 1926 and 1946 branch factories accounted for 54.5 per cent of the increase in manufacturing employment in the Pacific southeast area. Although branch plants in Los Angeles represent only 15 per cent of the new plants, it is estimated that they provide from 40 to 45 per cent of the new industrial employment. Of some $63,458,000 invested in manufacturing facilities in Los Angeles in 1948, one-fourth of this total was provided by 32 companies with national distribution which was new to the Los Angeles area. It has been estimated, also, that 50 per cent of the demand for industrial sites in San Francisco has come from established eastern firms.[4] To a degree that cannot be precisely determined, therefore, the control of industrial facilities has been shifting to national concerns. In the event of a depression, of course, branch plants and assembly plants would be closed or employment reduced before the parent plants were affected. The trend toward branch plants has gone so far, in fact, that one study refers to California as "a branch-plant empire."

Certain California communities, Los Angeles in particular, have also been paying an enormous price for industrial expansion in terms of the inroads that are being made on a basic resource, namely, food production. In the last ten years, manufacturing plants have taken over fully 50 per cent of the pasture lands formerly used for dairy herds in Los Angeles County, and the dairies are retreating as industry advances. The county now has 293,000 acres in crop production as compared to 311,000 acres in 1940—a factor of prime importance when one realizes that agriculture in urbanized Los Angeles County produced an income of $222,882,-990 in 1948. For many years, in fact, Los Angeles has been one of the richest agricultural counties in the nation. But, nowadays, the near-at-home food supply which has been such a factor in the city's expansion—Los Angeles has had to import far less food than other cities of comparable size—has been declining in the face of the constant encroachment on agricultural lands by industrial plants.

Then, again, the growth of industry has brought a host of prob-

[4] *Ibid.*, March 2, 1947.

lems some of which, of course, are peculiar to industry in California or have some special relevance here. "Symbolic of these (problems)," writes Dr. James J. Parsons, "is the low, gray pall of exhaust, incinerator, and factory fumes that lingers for most of the summer over the Los Angeles lowlands, trapped by the coastal temperature inversion and the mountain barriers to the north." For decades, now, the land-moving afternoon breeze from the ocean, and the mist which accompanies it, have been a major factor in keeping Los Angeles cool in the summer months. But, with the arrival of industry, the land breeze, once an asset, now serves to hold the fumes of industry over the city; fog, once hailed with relief in the summer months, has now become "smog," a major problem. Once the experts began studying this problem, however, they discovered, as one might have expected, that the "smog" problem in Los Angeles is unique. Los Angeles County is today the largest industrialized subtropical urban area in the world. It lies, also, in an enormous basin, ringed about with high mountains, and it is this combination of sub-tropical climate, mountain barriers, the land breeze, and the basinlike location that have made "smog" a unique problem in Los Angeles. Smog and fumes accumulate for days and remain trapped in this basin, unable to escape. No problem of the post-war period has occasioned more agitation and discussion in Los Angeles than the problem of "smog," which is far from being solved today.

The more one ponders certain aspects of the industrialization of California the more wisdom one can find in some comments by Dr. Parsons. "In concentrating its growth so heavily in California's two great metropolitan districts," he writes, "the West may well be building its house of cards. As the man-made superstructure of California's economy towers higher and higher, its dependence on the cooperation of a capricious Nature increases commensurately. The vulnerability of its water and power supplies alone suggests serious doubts as to the wisdom of continued expansion, at least until the economic application of atomic energy to peaceful ends. has been demonstrated. . . . In terms of its resource base California is probably being drastically oversold as a future industrial center. Yet the rising tide of 'progress,' defined as bigness in everything, is not likely to be stopped by less than war or earthquake, drought or economic collapse."

The casual reference in this statement to earthquakes, in terms of the concentration of population in large urban areas in California, is not to be lightly dismissed. California is an active earthquake country. It has in the San Andreas fault, along with its other wonders and marvels, the largest known earthquake fault in the world which extends more than a thousand miles and knifes through urban Los Angeles. Strange developments have been taking place along this fault, the meaning of which the seismologists do not pretend to understand but which nevertheless disturb them no end. California's earth is stretching itself slowly along a 600-mile section of the San Andreas fault. Records show that there has been a northwesterly shift of 10 feet in the earth markers along this fault in the last 63 years. Earth "dips" of four to five feet have also been recorded in the San Jose area and portions of the San Joaquin Valley. What the scientists do agree on, however, is the fact that, one of these days, California will experience another major earthquake. The newcomers, of course, dismiss the earthquake menace lightly and the chambers of commerce promptly mobilize reassuring statements whenever a scientist calls attention to the hazard.[5] But the older residents, those who have lived in California long enough to have experienced a major quake, speak of earthquakes with awe.[6]

TECHNOLOGY AND THE WEST

As a resident of Los Angeles, I was once deeply impressed by a remark of Morris Markey's. "It struck me as an odd thing," he said, "that here, in Los Angeles, alone of all the cities in America, there was no plausible answer to the question, 'Why did a town spring up here and why has it grown so big?' " Actually there is a

[5] Los Angeles *Daily News*, April 26, 1949.

[6] Although it is painful reading to a Californian, and is listed in none of the bibliographies, the reader might consult: *Earthquake History of the United States*, Part II, *The Stronger Earthquakes of California and Western Nevada*, by H. O. Wood and N. H. Heck. Serial No. 609, 1941, Coast and Geodetic Survey, U. S. Department of Commerce. The report lists 257 earthquakes between 1769 and 1940, of which 45, in the saturnine language of the seismologist, are "outstanding."

simple answer to this question—if one will look at a map. Today about four million people reside in the eight inter-mountain states: a population density of 4.8 persons per square mile as compared with 44.2 for the United States as a whole. Somewhere in the vast southwest there had to be *one*—not a dozen but *one*—major metropolitan area and Los Angeles is that area. Sparsity of population, usually a limitation on the growth of a city, is actually a major key to the expansion of Los Angeles. Within the Southwest, Los Angeles can never have any serious rivals. Hence, instead of being a fairly good-sized city, which it would have been if it were merely the "metropolis" of Southern California, it is destined to be one of the largest cities in the world. In the East and Middle West, one market area gradually merges into another; there are no intervening "blank" spaces. But since the Inter-Mountain West cannot support, within its boundaries, a truly major concentration of urban industry and metropolitan services, the whole area must be served by the west coast cities.

There is one thing the Inter-Mountain West has always possessed, and that is space. Curiously enough, space has come to be a great asset in this age of atomic power. Larger in area than the 10 southern states, the Inter-Mountain West had a population in 1940 that was about the size of that of North Carolina and not quite as large as the population of Los Angeles County. Yet, by a strange paradox, this area has a great economic and industrial potential. Today Los Alamos, New Mexico boasts the "best-equipped physics laboratory in the world," which already represents an investment of $500,000,000. Today 9,000 people live in Los Alamos and present plans call for accommodations to house 12,000 residents. Since the end of the war, Los Alamos has been transformed from a wartime "boom town" into a stabilized industrial community. "Remote and lonely on its 7,500-foot mesa," writes Hanson W. Baldwin, "Los Alamos stands today amidst snow-swept pines and cottonwoods, a strange symbol of the atomic age." Today, near the little town of Arco, Idaho, the Atomic Energy Commission has acquired a tract of 170,000 acres and is building another $500,000,-000 plant. In both cases, the remoteness of the areas and the superabundance of "space" were prime locational factors. Who knows just what these "symbols" mean in terms of the future development of the Inter-Mountain West? Who would have dreamed, a decade ago, that Los Alamos, New Mexico would ever boast of a half-billion dollar public investment, or that it would become one of the world's most important communities? What is the meaning of "Clementine," the only fast-reactor in the world utilizing fissionable materials, or of the new 12,000,000-electron volt accelerator, which has now been added to the Los Alamos plant?

At the present time "Clementine" may seem to be an asset of only hypothetical value to the West; but it may be that it symbolizes the equivalent of many times the energy of all the coal ever mined in Pennsylvania, of many times the energy generated at Niagara. The point to be noted is that "Clementine" is located in the heart of the Inter-Mountain West which is today the key area for the development of atomic power and the principal source of fissionable materials. Already uranium, the basic ingredient of atomic power, is scheduled to flow from five plants in Utah and Colorado by the end of 1949. How many plants and mills of this kind will there be, in this vast region, two decades from now, and how many people will they employ? This is the vast potential which gives promise, at long last, of transforming what was long regarded as the least productive into one of the most productive areas in the United States. It may well be, also, that the west coast has in this vast Inter-Mountain region a hinterland of incomparable wealth and purchasing power.

14

LADDER TO THE STARS

◇◇

ON JUNE 3, 1948, a thousand distinguished scientists and other dignitaries assembled on top of a granite mountain 130 miles southeast of Pasadena to take part in ceremonies dedicating the Hale Telescope: "the mightiest astronomical eye ever constructed by man for extending his power to explore the vastness of his cosmos." [1] This new "ladder of light" and "tower to the infinite," with its 200-inch, 15-ton mirror, now makes it possible for scientists to see and photograph island universes one billion light-years from the earth. "Climbing to the uppermost rung of his new 'Jacob's Ladder' to the stars," writes W. L. Laurence, "man will be able to solve many a cosmic mystery now beyond his reach." The Rockefeller Foundation alone has contributed more than $6,500,000 toward the cost of constructing the telescope and the Palomar Observatory.

Coincident with the dedication of the Palomar Observatory, an attentive newspaper reader might have noticed some related items in the California press. For example, on May 2nd, 1948, the Rockefeller Foundation announced that it had just made a $700,-000 grant to the California Institute of Technology for long-range research in biology and chemistry. Or the newspaper reader might have read on April 27th that the University of California was building a $9,000,000, 184-inch cyclotron, capable of whipping atomic bullets around a course until they reach speeds in excess of 60,000 miles a second with a striking force ranging from 6 to 10 billion volts. Just as the Hale Telescope at Palomar is today the most powerful instrument of its kind in the world, so the new cyclotron will be the world's largest and most powerful atom-smasher.

[1] W. L. Laurence, *New York Times*, June 3, 1948.

The same reader might have noticed that the University of California is also building a new $985,000 radiation laboratory and that Stanford University is building a 160-foot linear accelerator or atom-smasher which will hurl atomic bullets with the force of one billion electron volts. He might also have read, about the same time, that mesons—mighty atomic particles which promise to lead science into a great new domain of atomic energy—were produced for the first time in history in a laboratory at the University of California. Or his attention might have been attracted by the story on September 17, 1948 of how Drs. Isadore Perlman and R. H. Goeckerman of the radiation laboratory of the University of California have discovered a process called "fast fission"—the most violent atomic explosion yet known.

Reading these items, the same person might have recalled certain names which have brought world-wide recognition to American science—Robert A. Millikan, Thomas H. Morgan, Ernest O. Lawrence, Carl D. Anderson. These men were all associated with California institutions, and are all Nobel prize-winners. Intrigued by this pre-eminence of California institutions in basic scientific research, our reader might then have consulted American Men of Science and been impressed by the number of men who have made outstanding scientific achievements—who are associated with California colleges and universities. Of 65 scientists awarded the Medal of Merit for their work with the Office of Scientific Research and Development during the war, nine were members of the faculty of the California Institute of Technology.[2]

What is there, then, about California that has brought about this remarkable concentration of scientific talent? There is, of course, always an element of "chance" or "luck" in scientific research; often a particular discovery just "happens" to be made at a certain place at a certain time. But the concentration of a series of discoveries in a particular area within a remarkably brief time-span gives rise to an inference that there must be some underlying reason or explanation. "Chance" alone does not account for the con-

[2] See: *Scientists Starred, 1903–1943, in American Men of Science,* by Stephen Sargent Visher. John Hopkins Press, 1947. In 1938 Cal Tech had 23 men newly starred in *American Men of Science.*

temporary pre-eminence of California in basic scientific research; nor does it explain why the University of California is today not only the world's largest university but, in terms of its many-sided scientific developments, perhaps the outstanding university in the modern world. But there is an explanation—at least a partial explanation—for both developments and it is to be found, as one might expect, in the exceptional environment of California.

CERTAIN CALIFORNIA ECCENTRICS

Perhaps the point at which to begin the story of California's rise to scientific eminence is with James Lick, the eccentric San Francisco millionaire. Born in Germantown, Pennsylvania in 1796, Lick was a carpenter and joiner, later a piano-maker, who spent some 17 years in South America where he accumulated a modest fortune. He arrived in San Francisco in 1847, on the *Lady Adams*, with $30,000 which he proceeded to invest in real estate on the eve of the discovery of gold. At one time he owned Lake Tahoe and Santa Catalina Island. He it was who built in San Francisco one of the most famous hotels of the early West: the celebrated Lick House, with its floors of rare inlaid costly woods which he insisted should be kept as highly polished as a piano case. A recluse, careless about his appearance, he had few friends, and rarely entertained. Lick was something of a free-thinker; among his benefactions was a sizable grant to the Thomas Paine Memorial Society in Boston. During the gold rush period, Lick built a flour mill at San Jose at a cost of $200,000 which was one of the wonders of California. Known as "Lick's Folly," the mill was made of imported mahogany and other rare and costly woods. The mill was surrounded by carefully laid-out orchards and gardens for which Lick imported trees, plants, and shrubs from the four corners of the earth. In addition to subsidizing the Thomas Paine Memorial Society, Lick provided free baths for the residents of San Francisco, established an old ladies' home, an orphanage, and gave large sums to the Society for the Prevention of Cruelty to Animals.

No one seems to know just how, when, or why James Lick became interested in stars. But, shortly before his death, he expressed to Professor Davidson of the University of California a desire to endow an observatory which should have "a telescope superior to and more powerful than any telescope ever made." The observatory was originally planned for a site at the corner of Fourth and Market Streets in San Francisco but Lick had the idea that he wanted it located on a mountain top, and sites at Lake Tahoe and Mt. Helena were considered before the observatory was finally located on Mt. Hamilton, a mile-long ridge near San Jose, 4,029 feet in altitude, with a commanding view of the Santa Clara Valley. This site appealed to Lick, among other reasons, because it overlooked his former home and the location of the famous flour mill. One year before his death in 1875, Lick turned over $700,000 to the University of California to build the observatory. As a condition to making the gift, he insisted that the county build a road to the observatory site. Long before the road was completed, however, supplies and equipment were freighted up a rough trail to the summit, a five-day round trip. Lick Observatory was finally completed in June, 1888. Eighteen months earlier, the body of James Lick had been brought from the Masonic Cemetery in San Francisco, and placed beneath the pier which now supports the main telescope.

Lick Observatory was the first great mountain observatory. Previously, little thought had been given to the location of astronomical observatories, most of which had been adjuncts to universities and colleges. For whatever reason, Lick had the idea that observatory sites should be selected in reference to the specific advantages offered by particular sites and he was the first, so to speak, "to go to the mountains" in quest of a site. For many years, Lick Observatory had one of the largest telescopes in the world; in fact as late as 1935 its 35-inch telescope was "the second largest refracting telescope" in the world. Many important discoveries have been made at the Lick Observatory. In 1892 the fifth satellite of Jupiter was located (the first to be discovered since the time of Galileo) and Lick astronomers took the first successful photographs of comets and the Milky Way. The modern study of nebulae was begun at Lick in 1898, and since then more than 33 comets and 4,800 double stars have been discovered and charted at the observatory.

Lick Observatory was quite an acquisition for an institution like the University of California which, in the 1880's, was merely a small state university with little academic, and virtually no scientific, distinction.

Still another California eccentric who played a role in the development of mountain observatories, prior to Hale's work, was Thaddeus Sobreski Coulincourt Lowe. During the Civil War, "Professor" Lowe made one of the first balloon ascensions in America. On coming to Los Angeles in 1888, to build one of the first gas plants in California, Lowe became interested in the scenic resources and scientific possibilities of the mountain range which towers behind Pasadena. He it was who built the Mt. Lowe Scenic Railway, a cable line that ran from a point near Pasadena to the top of Mt. Lowe. Once the line was completed in 1893, Lowe proceeded to build the Chalet or Alpine Inn and the Echo Mountain House on the top of the mountain. And it was here, also, that he built the Mt. Lowe Astronomical Observatory in 1894 which functioned for some years under the direction of Dr. Lewis Spence. Lowe was obsessed with the idea that a great institution for the study of "pure science" might one day come into being on the mountain top in connection with the observatory. At an even earlier date, however, Edward F. Spence had designed an observatory for Mt. Wilson and had announced that he was prepared to give properties valued at $50,000 to the University of Southern California for the establishment of an observatory on Mt. Wilson. Spence, a Southern California booster, had become interested in the Mt. Wilson project as a means of offsetting the value of Lick Observatory as a tourist attraction. If northern California had an observatory, then Southern California must also have one. The project failed because Southern California did not have an eccentric millionaire to put up the necessary funds for its realization. When the project collapsed, however, Harvard University investigated the site but President Charles W. Eliot concluded that the scarcity of water made the site impractical. Before the project was abandoned, however, Spence had ordered a 40-inch object glass from the famous glass-grinders, Alvan Clark & Sons. As a matter of fact, it was this glass that George Ellery Hale secured for the telescope at

Yerkes Observatory, and for a time this was the world's largest telescope. For ten years after Spence's project collapsed, Mt. Wilson remained a wilderness but it had been "discovered," scientifically, and interest in the site had been aroused.

THE DEVELOPMENT OF MT. WILSON

It was the eccentrics, Lick and Lowe, who first stimulated an interest in mountain observatory sites in California but it was a great scientist, George Ellery Hale, who first realized the possibilities of such sites for scientific research. Hale was the son of a wealthy Chicago manufacturer of elevators and other types of hydraulic equipment. Upon his graduation from Massachusetts Institute of Technology, he visited Lick Observatory—on his honeymoon—and became convinced of the importance of mountain sites. He came to realize that the spectroscope and the photographic plate were the coming instruments of astronomical research. Hale had long wanted to combine "physics and chemistry with astronomy" in a new scientific discipline. He was later to found, in 1895, the *Astrophysical Journal*. Greatly impressed by the 36-inch telescope at Lick, and hearing of the 40-inch telescope which had been ordered for Mt. Wilson, Hale induced Charles T. Yerkes, the Chicago tractor king, to furnish the funds for the construction of Yerkes Observatory which was completed in 1897 as part of the University of Chicago. The son of a rich man, Hale knew how to approach the rich for funds; in fact it has been said that he was one of the first American scientists to enlist the aid of large private fortunes in the advancement of basic scientific research.

Not content with having built the "world's largest telescope" at Yerkes Observatory, Hale wanted to build a still larger telescope and, this time, he determined to select the site with the greatest care. In 1903 he succeeded in interesting the Carnegie Foundation and, from this source, obtained the funds to send his colleague, Hussey, on a site-exploring expedition. After traveling throughout western America and Australia, Hussey finally narrowed the possible sites down to five, two of which were located in Southern California—Mt. Wilson and Mt. Palomar. Hale then came out, looked over the two sites, and concluded that Mt. Wilson was the better. On this trip he also succeeded in inducing a wealthy Los Angeles businessman, John D. Hooker, to put up a thousand dollars so that he might bring camera equipment to Mt. Wilson to test its fitness as an observatory site. On the basis of these photographs, he was then able to convince the foundation that it should finance the construction of a new observatory.

Although Lick was the first mountain observatory, its location was based on the whim and caprice of an eccentric millionaire. But Hale's conscious search for the best possible site was an exceedingly novel idea. "His idea," writes G. Edward Pendray, "of placing an observatory where the seeing was best, though obvious now, was novel and almost radical in 1904. Previously most observatories had been a part of a college or university, and the sites were chosen with a view to making an impressive showing among the university buildings." It is important to note, therefore, that Mt. Wilson was the first of the world's great observatories which was selected only after a careful survey had been made of other possible sites. Furthermore, Mt. Wilson was consciously planned as *a great observatory,* designed for "the exploration of unfamiliar fields." From the first, it was Hale's conception that Mt. Wilson should "contribute in the highest degree possible to the solution of the problem of stellar evolution."

Now the question arises, of course, as to why Hussey and Hale finally selected Mt. Wilson. Once they had hit upon the idea of consciously selecting the best site, certain natural limitations as to latitude and altitude narrowed the possible range of desirable sites for a large telescope. "If the observatory," writes Pendray, "is located too far north, the bulge of the earth will cut off important areas across the celestial equator which can be observed from stations near the middle of the temperate zone. On the other hand, if it is too far south, the circumpolar stars will not rise far enough above the horizon." Approximately three-quarters of the entire celestial sphere can be observed between the latitudes of 30 and 35 degrees north. Mt. Wilson is in this region but so are many other areas in California and in other states. Why, then, was Mt. Wilson finally selected?

The specific factors which dictated the selection of Mt. Wilson as the site for the new observatory are numerous and, considered in the aggregate, they make a most unique combination. In the first place, height is important (Mt. Wilson is 6,000 feet); that is, certain elevations make for greater visibility. But height, per se, is not the decisive factor. For example, Hale discovered that Pike's Peak is a most undesirable site because snow, electrical storms, and low temperatures interfere with observation. Since the dome of an observatory cannot be heated, great altitudes must be avoided. What is needed is an area of a certain elevation, relatively free from rain, with long periods of unbroken weather, and with little overcast or clouds. A good observatory site must also be fairly remote and isolated so as to get away from the smoke and grime of cities; but it must be near a center of population from which supplies can be secured, where machine shops are located, and, also, from which visitors can be brought back and forth without too much difficulty. One of the factors in Hale's selection of Mt. Wilson, as he put it, was "the possibility of establishing the shops, laboratories, and offices in the City of Pasadena, within easy reach of large foundries, supply houses, and sources of electric light and power."

Located 30 miles from the Pacific, Mt. Wilson was free of fog and clouds; it was high but not too high. It had what astronomers call "an optically homogeneous" atmosphere. Once a road had been constructed to the summit, it was within a few hours traveling time from the offices and shops in Pasadena. Here it was possible to photograph the sun 300 days in the year; during one season, photographs were made on 113 successive days. The air was clear and steady, without heat-waves or the excessively bright sparkle at night which often interferes with astronomical observation. For weeks at a time, the sun and stars blazed without a wisp of cloud, and there was little wind. The daily range of temperature was so small that "atmospheric wobble" was negligible. The dome of the telescope did not need to be heated in winter, and, even during the rainy season, the air had a wonderful transparency. All of these factors enter into the selection of observatory sites but they have a special relevance to the locations for large instruments. Certain of these factors existed in reference to the other sites considered but it was the unique combination, in one site, that brought about the selection of Mt. Wilson. Here, once again, California was the great exception.

With his wonderful ability to manipulate millionaires, Hale was able to get large sums for the construction of the famous 100-inch telescope at Mt. Wilson from the retired tycoons who lived in Pasadena. John D. Hooker, who put up the money for the original survey of the site, gave $45,000 toward the cost of the project. In 1917 the great telescope at Mt. Wilson was first trained on the stars and it remained, for twenty years, the world's largest telescope. It was not long before the Mt. Wilson scientists issued their famous *Catalogue of Selected Areas,* giving the magnitude, on a uniform scale, of 67,941 stars. Here was assembled, according to Pendray, "one of the finest staffs of professional astronomers anywhere on earth" with wonderful facilities including a technical library of more than 13,000 volumes and 10,000 monographs and bulletins. In Pasadena, Hale built a large laboratory and optical shop where invaluable experimental work has been done in optical research and the technical phases of telescope and observatory construction. It was at Mt. Wilson that Albert A. Michelson did some of his most important work and here Eddington, Jeans, Einstein, and many world-famous scientists came to confer with Edwin Hubble, Walter S. Adams, and the other members of the Mt. Wilson staff, and to acquire the data used in their scientific formulations. For many years, Mt. Wilson has furnished "the bulk of the mountain output of cosmological data" used by astronomers, physicists, and chemists throughout the world. It would take a volume merely to list the accomplishments, major and minor, which have been recorded at Mt. Wilson. Incidentally it might be noted that since 1917 Mt. Wilson has been regularly visited by more than 75,000 sight-seers each year.

NOT THE MOON

Never satisfied with his achievements, George Ellery Hale had no sooner built the Mt. Wilson telescope than he began to lay plans for the construction of a still larger instrument. In the late twenties, he managed to obtain a commitment from the Rockefeller Institute to assist in financing the 200-inch telescope now in operation at Mt. Palomar. The grant which the institute made for this project, $6,000,000, is the largest single grant it has ever made. The story of the construction of the "glass giant of Palomar" is one of the most dramatic stories of our time. A story of incredible hardships overcome, of the dogged patience and courage of the scientists who worked on the project, of amazing technical resourcefulness. After years of work, the giant glass arrived in Pasadena on April 10, 1936. The story of how the glass was transported across the continent, as told by David O. Woodbury, is a saga in itself. The Palomar telescope is today the world's largest instrument for the enhancement of knowledge (the observatory shell itself is higher than a 15-story building). "Not the moon," writes Woodbury, "but the very boundaries of space" are the objectives upon which this instrument will be trained. It is, as Dr. Max Mason has said, "the springboard from which the philosopher may leap ahead toward the solution of the problem of the universe."

The checking, testing, and surveying that went into the selection of Mt. Wilson as an observatory site was almost superficial by comparison with the preliminary investigations that preceded the selection of Mt. Palomar as a site for the "glass giant." The most elaborate investigations, of every dimension, conducted by many scientific disciplines, were made in locating a site for the 200-inch telescope. At the time the Mt. Wilson observatory was completed, some 330,000 "lights" glittered on the coastal plain at its base but, by the time the 200-inch telescope was projected, the number of lights had increased to 2,500,000. For this reason alone, some other site had to be found for the giant. Every factor which had entered into the selection of Mt. Wilson also played a part in the selection of Palomar. Palomar is isolated but still accessible: a hundred miles from Los Angeles; fifty miles from San Diego. It has an abundant water supply; there is no possibility of the encroachment of towns or factories within a distance that would interfere with observation; and the observatory itself is located on a solid granite rock formation so that the earthquake hazard is negligible. To house the giant, the technicians needed a small flat place of a certain height (about 6,000 feet), "small enough to avoid storing much heat, large enough to be out of the reach of the updrafts from canyon 'chimneys'; high enough to take advantage of the steady upper air." As a site, Palomar has every advantage that Mt. Wilson possessed plus some advantages of its own. It is not only one of the most important centers of scientific research in the world, but it is well on the way to becoming one of the most popular showplaces in the United States, a magnet that will attract thousands of people to Southern California for many years.

It would be difficult to exaggerate the importance of the Lick, Mt. Wilson, and Palomar observatories in the furtherance of basic scientific research in California. To most people, astronomical research seems utterly remote from the mundane issues with which they are immediately concerned; but, in point of fact, this impression is entirely erroneous. Modern astronomical research has had a most important relevance to research in physics, chemistry, biology, geology, and many other fields. The implications of astronomical research in terms of other scientific disciplines are so great, indeed, that a layman cannot grasp, much less attempt to relate or explain, these implications as they bear upon general issues of concern to modern science. Suffice it to say that the findings which have been made at California's three great observatories have had a great deal to do with the pre-eminence of such institutions as the University of California and the California Institute of Technology in basic scientific research. It is not by chance, therefore, that such names as Millikan, Lawrence, Morgan, Anderson, Oppenheimer, Pauling, Seaborg, and Lauritzen, and a long list of similarly "starred" names from *American Men of Science*, are identified with one or another of these institutions.

SCIENCE IN AN ORANGE GROVE

In 1891 one Amos G. Throop founded an institution, located in a Pasadena orange grove, known as Throop University (the name was later changed to Throop Polytechnic Institute). Only by virtue of the most lax definition, could Throop be called either a "university" or an "institute." In the three small buildings located in the orange grove, courses were given in manual training, domestic science, sewing, knitting, and similar uncomplicated disciplines. For two decades Throop continued along these lines as the business section of Pasadena began to encroach on the orange grove. But when George Ellery Hale came to Pasadena to build the Mt. Wilson observatory, he decided that Throop Polytechnic might be transformed, in connection with Mt. Wilson, into a first-rate scientific institute. Ever the incomparable promoter, idea and action were invariably simultaneous with Hale and no sooner had he hit upon the idea than he arranged to take over the institute.

Fortunately for the future of American science, Orange Grove Avenue in Pasadena was lined in 1910 with the mansions of millionaires. After his experience in extracting money from the crabbed, tight-fisted Yerkes, the Pasadena millionaires were as putty in the hands of Dr. Hale. It was these men—Arthur H. Fleming, Norman Bridge, Henry M. Robinson, James A. Culbertson, Charles W. Gates, and others—who put up the money which made it possible for Hale to take over Throop Polytechnic, sewing classes and all. Fleming gave the new institution twenty-two acres of land which, with eight acres added at a later date, comprise the present campus. In 1913 Dr. A. A. Noyes joined the faculty on a part-time basis and, in 1917, Dr. Robert A. Millikan first became identified with the institution. It was Hale who finally persuaded Millikan to join forces with him in making the California Institute of Technology (the new name was adopted in 1920) a great scientific research institute. It was in 1920, also, that Arthur F. Fleming agreed to give the new institute his personal fortune as a permanent endowment.

It is doubtful if any parallel exists in the history of institutions of higher learning in America for the rapid strides which Cal Tech has made under Millikan's direction. In 1921 Cal Tech was hardly known in scientific circles; today it is one of the greatest institutions of its kind in the world. Dr. Millikan had two outstanding qualities as a director of this type of institution: a remarkable "feel" for what is relevant in the field of research—an almost intuitive awareness of the subject-matter that is most likely to pay the highest scientific dividends; and a genius for the organization and direction of research. Like George Ellery Hale, he could have been a successful businessman and he was certainly second only to Hale as a fund-raiser. Shortly after he took over the presidency of Cal Tech, he organized sixty Southern California millionaires into the California Institute Associates at a meeting in the home of Henry E. Huntington. Each of these men contributed a minimum of one thousand dollars annually for the privilege of belonging to the Associates and many of them contributed vastly larger sums. The availability of this "surplus" wealth largely made possible the rapid expansion of facilities at Cal Tech.

In analyzing the sudden emergence of Cal Tech as a major institution for scientific research, one starts with two obvious factors: the existence of exceptional leadership as represented by men like Hale, Millikan, and Noyes; and, second, the uniqueness of Mt. Wilson as an excellent site for astronomical observation. Had there been no Mt. Wilson, there would have been no Cal Tech. A third factor is the existence of "millionaire row" on Orange Grove Avenue. This was not, in the usual sense, another "millionaire's row," for most of the Pasadena millionaires were retired—their wealth was "surplus," idle, liquid wealth. For the most part, the millionaires on Orange Grove Avenue were men who had made their fortunes elsewhere and had retired in Pasadena, attracted there by the climate and surroundings, the scent of orange blossoms and the view of the mountains. The wealth which these men possessed was "settled," and, in some cases, "inherited" wealth; wealth which was not directly involved in industry. Furthermore, there has always been an aura of "culture" about Pasadena which found reflection in a certain "softness" and "mellowness" about these millionaires—grown lax in the sunshine—that Hale and Millikan manipulated to advantage.

A fourth factor might be said to relate to the intangible "spirit of the place." Enterprising, booster-made Los Angeles wanted an observatory to match the Lick Observatory in northern California. The observatories projected by both Lowe and Spence were essentially by-products of the first great real estate boom in the eighties. It was the booster spirit that prompted the Los Angeles businessman, John D. Hooker, to make the first important contributions to the Mt. Wilson project. The same spirit has had much to do with the success of Cal Tech's various fund-raising campaigns.

"Climate" accounts for the presence of millionaires in Pasadena, as it also accounts for Mt. Wilson; but what of the faculty? How was this small, almost completely unknown institution, located in far-away "crazy" Southern California able to attract, with inadequate funds, some of the most talented men in American science? Consider the case of Dr. Eric Temple Bell, one of the outstanding teachers of mathematics in the United States. Prior to joining the faculty at Cal Tech in 1926, Dr. Bell was teaching at Columbia University. A former Californian, he wanted desperately to return to the coast: one winter in New York was enough for him. In joining the Cal Tech faculty, however, Dr. Bell took a 50 per cent cut in salary which was only partially offset, at the time, by lower living costs in California. Over the years, Cal Tech has gradually made up a portion of this difference by salary increases; but Dr. Bell informs me that, on the basis of his salary at Columbia in 1926, Cal Tech would still owe him 16 per cent of his earning capacity between 1926 and the present time if it were to pay him, dollar for dollar, what he might have earned in the East. Nor is this an isolated case; many members of the faculty at Cal Tech came to Pasadena for precisely the same reason, namely, they liked living in Southern California. In this sense, therefore, "climate" has subsidized Cal Tech to the tune of many millions of dollars.

By charting fertile fields for research at Cal Tech, Hale and Millikan made it possible for the institution to reap a rich latter-day harvest in prizes, honors, and prestige. Since Cal Tech was deeply involved in rocket research as early as 1936, the government naturally centered all research of this type in Pasadena. Through the Office of Scientific Research and Development, contracts totaling more than $80,000,000 were awarded to Cal Tech during the war. This sum, furthermore, represents merely a portion of the money that has been spent, and is being spent, by the government in financing research at Cal Tech. Under the circumstances, therefore, it will be very difficult for other institutions to overcome the headstart which Cal Tech now enjoys in many fields of research.

It is apparent, therefore, that Cal Tech has been a major asset to Southern California. But to calculate the value of this asset one would have to investigate the role that research in aeronautics and aerodynamics at Cal Tech has played in the development of the aviation industry in Southern California. Cal Tech is the home of the Guggenheim Aeronautical Laboratory and of the Cooperative Wind Tunnel Project, financed to the tune of $2,500,000 by the aircraft industry in Southern California. One would also have to know something about the role that geological research at Cal Tech has played in the development of the oil industry. In physics, chemistry, geology, climate-research, biochemistry, biology (Cal Tech recently received a $700,000 grant from the Rockefeller Foundation for research in biology), and a dozen other fields, the findings of Cal Tech scientists have been of enormous importance to the development of Southern California. It should be emphasized, again, that the resources of California have always been of a character requiring a high level of technological development to unlock their riches. It is only in relation to this background that one can sense, without being able to define, the vital role which basic scientific research has played in the development of the entire state. In this as in so many other fields of inquiry about California, one comes back to the "climate" and the exceptional character of the environment. Like many other California institutions, Cal Tech is, in no small measure, a product of the "climate."

Antennas on a Mountain Top

The value of Mt. Wilson to Southern California has not been restricted to scientific research; part of its value consists in its unique geographical relation to the rest of Southern California. When radio engineers began to investigate possible television sites in Southern California, they soon discovered that Mt. Wilson was ideally located for television transmission. For Mt. Wilson commands a "line-of-sight" view which makes possible an unusually clear reception as far north as Santa Barbara (100 miles), as far south as San Diego (100 miles), as far west as Santa Catalina Island (50 miles to the west and 22 miles offshore), and as far east as San Bernardino. From this 6,000-foot peak overlooking the coastal plain, there are few intervening ranges or other obstructions to direct, clear reception of an unusually high quality. Today on a short half-mile of hogback near the observatory are seven sites for television broadcasting stations, all in operation, representing an investment of $3,000,000. Unlike New York and other major broadcasting centers, where television stations are not closely grouped, Southern California television fans need not adjust their antenna to get any one of the seven stations now in service. There are relatively few potential program origination points in Los Angeles from which 6,000-foot Mt. Wilson cannot be seen. Programs originating in Hollywood are beamed to the Mt. Wilson transmitter by means of low-power, highly directional radio transmitters, and are then radiated from the antenna on Mt. Wilson in all directions. Already there is more concentrated television activity at Mt. Wilson than in any area of comparable size in the world and the unique effectiveness of this site as a transmitting center is likely to make Los Angeles the center of television for America.

With seven television stations in service, Los Angeles has already become in relation to television what it has for many years been in relation to radio, namely, an important center for the origination of programs. Many small local businesses have come into being to serve the new industry, including companies making coils, tuners, transformers, metal assemblies, plastic knobs, insulators, condensers, dials, grill cloth and glass parts, as well as sets. Los Angeles will probably produce the bulk of the film for national television programs, for it has the know-how, the theatrical talent, and the technicians. Hollywood's background in radio and motion pictures makes it, in fact, the logical center for reconversion to television. Although this development will have to mark time until more coaxial cables are built, Mark Woods of ABC, Harold J. Bock of NBC, and other network executives are agreed that Los Angeles is likely to become the television capital of the world. If it does, then the exceptional relation of Mt. Wilson to the rest of

Southern California and certain "freakish" aspects of the environment will be largely responsible.

With an estimated 4,000,000 people within the range of the transmitters on Mt. Wilson, only about 170,000 are known to be in so-called "blind areas." As a matter of fact, Southern California, according to the U.S. Navy Electronics Laboratory, enjoys three to four times the normal "line-of-sight" television range available to the rest of the country. Due to a "peculiar" quality of the environment, Mt. Wilson actually has a "line-of-sight" even in communities set apart by intervening mountain ranges. Television micro-waves travel at the speed of light. If they hit a surface that is the proper size in relation to the wave length, they will "bounce" off instead of being intercepted. The mountains separating the Mt. Wilson transmitters from such communities as Santa Barbara, San Diego, and Bakersfield, just happen to be of this character. Another peculiarity of the environment, namely, temperature inversion—a layer of warm air above the cooler surface air—also tends to "bend" the micro-waves, making for excellent reception over an unusually wide area. Temperature inversion, the same phenomenon that has created an "unusual" smog problem in Southern California, is an exceptionally important asset in relation to television.

The Right Side of the Mountains

The University of California is today the world's largest university, with 50 colleges, schools, and institutes, 2,000 faculty members, and a library of 1,500,000 volumes (the third ranking library in the nation). It is not the size of the university, however, that accounts for its fame today but rather its many-sided specialization in scientific research.[2] Curiously enough, the University of California owes a great deal, in this respect, to its traditional rival, Stanford University. Founded in 1891, Stanford was one of the first west coast universities to emphasize the importance of instruction *and* research. Under the leadership of David Starr Jordan, Stanford got off to a quick start, and soon outdistanced Berkeley as a research center. In the last twenty years, however, the re-

[2] See: *Fortune*, June 1946, p. 150.

lationship has been reversed. There had always been a special emphasis on science at Berkeley. Founded in 1868, the university came into being at a time when, to quote *Fortune* magazine, "the natural sciences had appeared in full brilliance." The youth of the institution and its remoteness from the older centers of learning "cut it off from the traditional New England insistence on the classics and the humanities." Few universities, it has been said, have taken greater pains than the University of California to get as many sciences into the heads of as many students. It was with the appearance on the campus of an outstanding scientist, G. N. Lewis, that the university really began to forge ahead in research, despite the early emphasis on science.

Berkeley, however, is only one of several west coast universities. Why is it that this one state university has been responsible for so many scientific discoveries? Today Berkeley is a great center of scientific research, whereas the University of Washington and the University of Oregon have failed to attain a comparable distinction. Part of the answer is to be found in the fact that in Oregon and Washington the basic sciences were divorced, at an early date, from the state universities and assigned to the agricultural schools: Washington State at Pullman and Oregon State at Corvallis. Washington is really two states, for the great farming areas "east of the Cascades" have little in common with the western, coastal part, of the state. The rivalry between these two sections has always been sharp and it was the political power of the eastern farming areas that brought about the assignment of the basic sciences to Pullman.

California also has an "eastern slope" but the "east of the Sierras" portion of the state is wholly insignificant in terms of resources, population, and political power. There are distinct subregions in California but essentially the state is one: a region in itself. By and large, the boundaries of the other Western states make little sense. As Dr. A. N. Holcombe has pointed out, the eight inter-mountain states "possess boundaries drawn without any regard to the interests of the inhabitants or the conveniences of the state governments." But California, for all its diversity, is a natural administrative unit and its boundaries do make sense. If the great Central Valley were located on the eastern side of the Sierras, California would almost certainly have had two state universities. But since the state is essentially one, California has been able to concentrate its educational facilities and to integrate scientific research to a remarkable degree. There is an agricultural school at Davis but it is an integral part of the state university, and the basic sciences have always been centered in the university proper. Comparison might be made, here, with Colorado, which has a state university, a college of agriculture, and a school of mines, all administered as separate institutions.

Today Berkeley occupies much the same relationship to the entire West that the University of Chicago occupies to the other schools and colleges of the Middle West. For years Chicago has been draining off the best men from the staffs of the state universities and private colleges in the area, as well as the cream of the graduate students. So far as important research in the basic sciences is concerned one can today jump from the University of Chicago to Berkeley. East of the Rockies the pull is toward Chicago; west of the Rockies the pull is toward Berkeley. In this as in many other respects, the entire West is California's hinterland. Berkeley's strength in scientific research, it has been pointed out, consists in its remarkable *specialization* and in the *diversity* of its achievements. One need only note, here, that California happens to have a highly diversified and specialized economy. Research has contributed to this diversity and specialization, just as the nature of the environment finds reflection in the type of research for which the state university is today world-famous. Berkeley has the world's greatest atom-smasher and it was here that meson was first made. It is here, also, according to David Lilienthal, "that some of the most exciting and significant research on earth is now going forward." [3]

By their remarkably successful emphasis on basic scientific research, both the University of California and the California Institute of Technology have a special relevance to the future of California. In a speech delivered on the occasion of a ceremony honoring his 80th birthday, Dr. Robert Millikan suggested the nature of this relevance. "Southern California," he said, "faces a

[3] *Life*, September 27, 1948, p. 118.

challenge. She has no coal, which has in general been considered the basis of the great industrial developments. In her semi-arid climate she has not the natural hinterland commonly considered essential for the support of such a huge population as wants to live here. In order to meet the *challenge* of these handicaps she must *of necessity* use more resourcefulness, more intelligence, more scientific and engineering brains than she would otherwise be called upon to use. . . . In a word, if Southern California is to continue to meet *the challenge of its environment* . . . its supreme need . . . is for the development here of men of resourcefulness, of scientific and engineering background and understanding—able, creative, highly endowed, highly trained men in science and its application" [emphasis mine]. In the last analysis, it has been the challenge of this unique environment which has brought forth the remarkable scientific advances that have been made in California in the last three decades.

[15]

THE ONE-LEGGED GIANT

◇◇◇

A STATE with many exceptional advantages, California also has some exceptional disadvantages, and of these the scarcity of water is certainly the most important. The water problem is not one but a dozen interrelated problems, affecting every aspect of the economy of the state. Problems, of course, have various dimensions; but the water problem in California is uniquely multidimensional with each dimension being most complexly interrelated with every other. In fact the word "problem" does not accurately describe this complex of issues. The water problem in California is an ecological anagram, a sociological maze, an economic logograph. Since some new dimension must be added to the word "problem" before it will carry the full weight of the facts, the first task is to suggest the nature of the problem itself.

CLIMATE WITHOUT WEATHER

California's exceptional climate is, of course, the state's major asset. Strong on climate, California is notoriously weak on weather. To borrow a figure of speech from Leverett Richards, the west coast is like a one-legged man: Washington represents the head and shoulders, Oregon the heart and trunk, and California the elongated leg. The calf of this one leg is the Central Valley; the heel is Southern California. For weather California must take what is left over from Washington and Oregon. The west coast's weather originates in the Gulf of Alaska, and from there sweeps south and east across the coasts of Oregon and Washington. "Only the outskirts of the bigger storms," writes Mr. Richards, "trail down the coast as far as California." Such storms as reach California are really the leavings, the remnants of rain.

As one moves from north to south along the west coast, the amount of precipitation gradually diminishes; the rains become thinner and thinner. From 20 inches of rainfall at Red Bluff and 17 inches at Sacramento, only the remnants of storms reach the southern end of the Central Valley landlocked by the 5,000-foot coastal range on the west and the 14,000-foot Sierras on the east. Fresno, in the heart of the valley, has an annual average rainfall of 9.35 inches; Bakersfield, in the ankle of the giant, gets about 5.50 inches. The Southern coastal plain, the giant's heel, gets about 7 inches, being shut off from the extremely dry southern San Joaquin Valley by the 6,000-foot Tehachapi Mountains. San Diego, in the extreme toe, averages about 4.5 inches annually. San Diego, in fact, is very close to the line of demarcation along which the North Pacific cyclonic storms cease to have an appreciable effect.

Not only do the North Pacific storms taper off as they sweep from north to south along the coast, but they are what might be called "migrating storms." From March to July, the winter rains retreat up the coast from South to North with the extreme northward migration being reached in late July. The one-inch line of rainfall on the coast is somewhat north of San Francisco in May but, by June, it has retreated to the northwestern California coast. Then, from August through December, the belt of winter rains starts moving southward once again. Each month of this period shows an extension of the area covered, and an increase in the amount of rain. Although there is little difference in the temperature along the coast, the amount of rainfall varies enormously. There are areas in the Olympic Mountains which have the heaviest annual recorded rainfall of any areas in the world but, once the Oregon-California boundary is crossed, the amount of rainfall decreases rapidly. In California, also, there is a great variation in rainfall on an east-to-west basis: from 50 inches on the coast to less than 10 inches on the Nevada desert.

But average figures fail to tell the story of the variation in rainfall in California for the amount of precipitation varies from year to year, from wet years to dry years, and between subregions in the state. Rainfall varies as much as 700 per cent from one year to the next. For example, in 1850 Sacramento recorded 36 inches of rain but only 4.71 inches were recorded in 1851. In coastal Southern California, the annual rainfall will often increase 10 inches in a single steep mile as one ascends from the foothills to the mountains. Riverside, 850 feet in elevation, has 10.8 inches of rainfall; San Bernardino, 10 miles distant, elevation, 1,050 feet, has 16.1 inches. Along the rim of the San Bernardino Mountains, rainfall runs up to 30, 40, and 50 inches annually. It should also be noted that the amount of the seasonal run-off varies enormously from year to year and this variation correlates, not only with the amount of snow and rain, but with variations in the temperature. Ice melts into 100 per cent water; mid-season snow will melt into 35 per cent water; but late snow will melt into only 18 or 20 per cent water. Similarly, rates of evaporation show amazing variations and in an inverse relation to the amount of annual rainfall.

It so happens, also, that the areas of relatively heavy rainfall in California are not located where they should be in relation to concentrations of population. Water is brought great distances to the coastal cities by huge engineering projects such as the Hetch-Hetchy and Mokelumne projects in the North and the Owens River and Colorado River projects in the South. With 1,400,000 acres of arable land (6 per cent of the state's total) and 45 per cent of the state's population, the Los Angeles Basin has only .06 per cent of the natural stream flow of water. Although nearly all the rainfall occurs in California between November 1 and March 1, years of heavy rainfall alternate with years of excessive drouth; nor does anyone know how much it will rain in any particular season or just when the rains will come. Yet, paradoxically, floods are, and always have been, a major hazard in the state!

California is, indeed, a one-legged giant. The one leg is climate; the missing leg is weather. With this background in mind, it will be interesting to see what happens in California when the rains fail to come.

THE DROUTH OF 1948

During the more than 100 years in which rainfall totals have been recorded in California, so-called "dry cycles" have recurred about every thirty years, and have lasted about seven years. The "dry cycles" do not commence with the year of least rainfall and, in fact, the inception of a "dry cycle" is scarcely noticeable for the first year or so. The cycles usually begin with a year of reduced snowfall in the mountains or of an exceptionally rapid run-off. California "dry cycles," however, are by no means uniform in duration or intensity.

On March 13, 1948, the rains finally came to break a drouth of sixty weeks, the worst of its kind that the state has experienced in nearly fifty years. In fact, December, 1945, was the last month of abundant rainfall, so that 1948 really marked the third year of a recurrent "dry cycle." The 1947 rainfall in Los Angeles was the lowest ever recorded being 50 per cent of the average or normal rainfall, and the run-off was estimated to be 75 per cent of "normal." Instead of noting the tell-tale symptoms of drouth which had become apparent in 1946, the state drifted along, gambling on rain, and hoping against hope that the weather would "break." Then, as always happens, the people awoke to the fact that a serious drouth was on, and the Governor, on February 21, 1948, proclaimed a state of emergency in 28 counties, embracing about two-thirds of the entire area of the state, appointed a "czar" to conserve power and ordered a 20 per cent reduction in the amount of power consumed. Before mild rains finally came in March, 1948, portions of California were as dry as the Gobi and Sahara Deserts. When the rains did come, moreover, it was a question of "too little" and "too late": not enough rain fell to fill the underground basins and storage reservoirs and the rains came too late to save many crops.

It is extremely difficult to measure the effects of a drouth in California for many of these effects do not become fully apparent for many years. A vineyard or orchard which is sun-burned one year may not show the full damage for two or three years. Although the tree-and-vine damage cannot be measured, the 1948 drouth struck a heavy blow at the cattle, sheep, dairy, and hay and grain industries. As the foothill pasture areas of the San Joaquin Valley began to "burn," livestock interests were forced to ship cattle outside the state for pasturage and to make "distress" sales on a glutted market. Over 200,000 head of cattle had to be shipped from drouth areas; dairy herds were sharply reduced; and 50,000 sheep were shipped outside the state for pasturage. Thousands of head of livestock, "too thin to stand the trip," were sold on the market. Great cattle interests, such as the Kern County Land Company, reduced their herds by approximately 50 per cent. By September 1948, combined losses were estimated to be close to $100,-000,000.

By March of 1948 the foothill ranges were "as clean and bare as the top of a freshly scoured kitchen range." With only a trickle of water left in the streams, gasoline engines and windmills were used to pump water for range cattle and, in some areas, water was hauled in by truck. Around Avenal and Kettleman City, headquarters for several big cattle companies, the range "looked as naked as a dance floor." "The saddest sound between the Diablos and the Santa Lucia Range," reported Stanton Delaplane in the San Francisco *Chronicle*, "is the rattle of the loading gates and the lonesome whistle of the cattle trains in the night. The four-and-a-half-million dollar cattle business of the King City area is moving out until the rains come again." Over 5,000 head of starving cattle were shipped from the 40,000-acre Santa Rosa Island to the mainland. Hardly a sprig of green was left on the island and the trees, one reporter noted, "looked sick, with their leaves hanging in parched dejection." Cattle were shipped to Oregon in such volume that the state inspection laws, designed to eliminate bovine diseases, could not be adequately enforced. Heavy shipments of livestock from drouth-stricken areas tied up rail transportation and brought a clamorous demand for lower freight rates. Noting the signs of a dry cycle, many farmers switched from alfalfa to cotton to conserve water and thereby reduced the amount of available feed for cattle and sheep. Hay and grain growers sustained losses estimated at $20,000,000. With the supply of hay reduced, prices jumped to $40 a ton thereby forcing many dairymen to sell or reduce their herds. Today the old-time cattlemen, those who remember back "four dry spells," are gambling on the proposition that the drouth will end, eventually; but only the giants among them are likely to survive.

Starting with the cattle, sheep, dairy, and hay-and-grain indus-

tries, the effects of the drouth began to spiral outward in ever-
widening circles and with multiple and cumulative consequences.
Agricultural costs shot up more than $6,000,000 in Southern Cali-
fornia as more power was required for pumping and the produce
and floral industries, in particular, were hard hit. The 20 per cent
reduction in power affected nearly every industry in the state. The
costs of city government, in scores of communities, rapidly increased as special measures were taken to counter the effects of the
drouth. Before the year was over, it was painfully apparent that
the drouth had affected nearly every segment and facet of the
economy of California. But, as previously noted, the long-range
consequences could not be measured.

GEOLOGICAL WATERS

One reason why the effects of drouth cannot be measured is
that they cannot be seen: the real drouth is underground. As
everyone knows, California's rivers run "upside down": it is the
underground flow that is important. When farmers can see even a
slight foam of water in the Salinas River, they know that underground waters are flowing in large volume. To these farmers a
trickle on the surface of a stream has the same significance that a
river full to its banks would have to a midwest farmer. The farther south in distance from Sacramento, the greater becomes the
reliance on underground waters. Underground storage is the most
effective form of water storage in an arid or semi-arid environment. Loss-by-evaporation is reduced to a minimum; storage basins
do not have to be constructed; and the water is stored in the areas
where it is needed. Underground waters make up the "water
bank" of California, an asset of inestimable value, yet the state
lacks any form of statutory law regulating the right to pump
underground waters.

In many areas of the state the underground waters were so
plentiful, at one time, that artesian wells could be brought in by
shallow borings. The first artesian well was bored in Southern
California in 1868, and by 1900, some 11,000 artesian wells were
flowing in the area. These first wells were drilled to a depth of

only 40 to 50 feet and no pumping was necessary for the artesian
flow sprayed water into the air. At the present time, however, the
artesian wells have long since ceased to flow, and more than 30,000
pumps tap the underground supply in Southern California. At one
time, also, a similar artesian belt extended from Stockton to Bakersfield in the San Joaquin Valley, an area which once had 551 flowing artesian wells. These wells, too, have long since ceased to flow
and today the Central Valley has more than 50,000 wells from
which water is pumped: the biggest concentration of irrigation
pumps in the world.

What these first artesian wells tapped, of course, was the annual
underground flow of "surplus" water. At one time, California had
89.9 per cent of all the farm acreage in America irrigated by artesian wells. However, since the demand for water increased more
rapidly than water resources could be expanded, the artesian flow
was soon exhausted. Wells were then drilled to a greater depth and
power was applied for pumping. The importance of this underground supply, in particular areas, can be suggested by the fact
that three-fourths of the irrigated farm acreage of Southern California relies upon water pumped from underground sources (water
imported from the Colorado and Owens Rivers is used for domestic and industrial purposes; not for irrigation). Large areas in the
Salinas and Santa Clara Valleys, and in portions of the San Joaquin Valley, are entirely dependent upon underground water
sources. Many towns and cities in California are likewise dependent upon the underground supply. San Francisco has over 150
wells in the downtown area, often in the basements of buildings,
which supply water for hotels, laundries, department stores, and
many buildings.

Over a period of years, the water level has been dropping in all
but two areas of the state and continues to sink lower and lower as
a result of excessive pumping. In the important farming area between Merced and Bakersfield, the overdraft on the underground
water supply is currently estimated to be in excess of 1,000,000
acre-feet a year.[1] In the Evergreen area of the Santa Clara Valley,
the water level has dropped 100 feet in the last decade. Since 1920,

[1] Fresno *Bee*, March 22, 1949.

the water level has dropped 60 feet in Kern County and 130 feet in the Lindsay area. In the twelve-month period ending in October, 1947, the ground water level dropped 11 feet in the Arvin-Edison district; 30 feet in portions of the Salinas Valley; and, in some areas, as in the Chula district near Monterey, the wells are already beginning to pump salt water. The drop in the underground water level in California is not to be explained by periodic drouths; drouths aggravate but they do not cause this continuous overdraft on the underground supplies. The fact is that more water has been taken out of the underground "bank," over a period of many years, than has been replaced by normal seepage. In many areas of the state today it is not the annual underground flow that is being tapped, but the geological waters, an irreplaceable asset.

The more the water table drops in California, the deeper go the wells; the faster underground waters are pumped; and the greater becomes the burden on the utility systems. Normally, pumping operations do not start, in many areas, until midsummer: rainfall and surface waters sufficing until this time. In 1948, however, farmers began pumping six months in advance of their normal schedule, from necessity in some cases, in others to get their supply of the underground water "while the getting was good." In March 2,500 farmers met in Fresno to pass sundry resolutions about conserving the water supply; but most of them hurried back to their farms to turn on the pumps weeks in advance of normal operations, knowing that their neighbors would do the same thing. In this frantic quest for water, hundreds of new wells were drilled, others were re-drilled and deepened, and wells already in operation were run "round the clock." In many cases, large corporate farming interests leased lands on the "west" side of the San Joaquin Valley, on wells in the middle of the Valley from which water is pumped to lease-holdings as large as 40,000 acres located ten, twenty, and thirty miles from the wells. Small farmers, in the area of these wells, invariably become concerned when they see that the local supply is being drained off to other areas and try to preserve the local supply by wasteful, badly timed, and incontinent pumping. In the period from 1947 to 1948, wells which had been dropping 10 and 20 feet a year, suddenly dropped 400 feet in six months. In areas of Kern County, the underground water level dropped so sharply that the land began to settle, buckling and severing pipes and canals and causing extensive damage to irrigation systems. On the "west" side of the Valley, hundreds of new wells were drilled to depths of 1,500 and 2,000 feet. Some of these deep-level wells were "pumping air" almost before they had been placed in operation.

What this means, of course, is that a resource of incalculable value is being extinguished. For pumping conducted on this scale and by these means is not pumping: it is "mining." It is almost impossible to measure the supply of underground deep-level water resources for the water is loosely held in sand and gravel deposits, but when the supply is exhausted it is exhausted for an indefinite time. These deep-level sources are geological waters which have been slowly accumulating since the days of the dinosaurs. Underground waters accumulate by a slow, steady process of seepage, with only a small portion of the surface supply, under normal conditions, sinking into the ground. Most of the surface water, in fact, either drains off or is lost through evaporation. Once depleted, these deep-level sources cannot be replaced within a period of time that would have any significance to persons now living. No one knows precisely how much water is still underground but it is known that in the lower San Joaquin Valley wells have struck granite rock and below this level there is no water. Underground reservoirs built up over a period of 400 to 500 years were exhausted in one year of excessive pumping in 1948. In the face of this sort of "run on the bank," it is apparent that the bank faces insolvency. It would be difficult to exaggerate what is at stake in this matter. The Central Valley is a larger and much richer agricultural area than the Nile Valley. It contains 59,000 farms, totalling 3,500,000 acres, and produces crops valued at more than a billion dollars a year—one half California's annual agricultural production.

The same factors that produce a water famine in California invariably produce a power crisis. The lack of rainfall, the late seasonal run-off, and the lighter snowfall reduced the storage capacity of power dams by 50 per cent in 1948. The more water pumped, the greater becomes the "load" on the utility systems; also, the deeper wells are drilled, the more power is required. In the Santa Clara Valley the demand for power increased 894.2 per cent in a single year (1947–1948), and the increase was 161.7 per cent in the San Joaquin Valley. At the same time, the water stored in the power-generating dams of the Pacific Gas & Electric Company, which was 61.6 per cent of capacity in 1947, fell to 44.5 per cent of capacity by 1948. Shortly after the first of the year, the power crisis reached a point where the state had to invoke daylight saving, appoint a "czar" to conserve power, and order a 20 per cent reduction in the use of power. The more power is used to pump water, the less power is available to industry.

Before 1948 was well-advanced, many farmers were operating their pumping plants by gas and diesel engines to conserve power. The strain on a utility system can only be sustained up to a certain power; if the demand is extended beyond this point, the engineers must "pull the switch" in order to save valuable machinery and equipment. On several occasions in 1948 it was touch-and-go whether the utilities could sustain the demand for power; hence the "brown-out" which spread throughout large areas of the state. In 1947 the canneries of California packed 69,562,041 cans of fruits and vegetables and the frozen food industry packed 87,569,-712 pounds of food. Both industries are as directly dependent upon an adequate water and power supply as any farmer in the state. Six ice plants in Salinas use 90,000 kilowatts of power 10 hours of each day for 10 months of the year. A 20 per cent cut in power for these plants means that 40 carloads of lettuce spoil every day. Thus the direct and indirect consequences of a water famine in California, if the famine is prolonged, can bring the entire economy of the state to a halt, for this economy is most intimately interrelated and interdependent.

THE INTRUSION OF SALT

Despite what has just been said about the calamitous effects of excessive pumping, and paradoxical as the statement may sound, the pumping of underground waters is, in many areas of the state, not only a useful but a necessary practice. "The ground water of arid regions," as Dr. Hans Jenny has pointed out, "in contrast to that of humid regions, is commonly rich in dissolved salts. If the ground-water table is near the surface, say within 5 or 10 feet, water tends to rise by capillarity and to evaporate, leaving the dissolved salts in the soil. In other words, the soil acts like a wick." There are thousands of acres of farm land in California, once highly productive, which have been severely injured by improper drainage and excessive irrigation. In many irrigated areas, it is absolutely vital that the ground water should be kept at a certain level through pumping; otherwise, the land becomes water-logged and alkaline. In the Turlock-Modesto area, a large part of the power available to the local irrigation district is used to keep the water level from rising. In these areas, irrigation waters are recaptured by pumping and used again and again. In fact, the recaptured waters make up about a third of the water supply. Pumping per se is not an unmitigated evil in California, but unregulated pumping is certainly evil. The absence of a state policy governing the use of underground waters is, in fact, the outstanding weakness in California's water conservation program.

Nor is the matter of excessive pumping related solely to agricultural areas, for many urban districts are directly dependent upon underground supplies for domestic and industrial use. The west basin of Los Angeles County, with 12 cities and an estimated population of 300,000, is largely dependent upon underground water sources. Today some 2,000 wells in this area are making a 100 per cent overdraft on the underground sources. That is, in excess of the annual replenishment through rainfall and other surface sources. In 1904 the water level, in this area, was from 20 to 40 feet *above* sea level. Today it is from 20 to 75 feet *below* sea level. As the water level has dropped, salt waters have intruded; at the present time the salt water intrusion is proceeding inland at the rate of from 500 to 1,500 feet a year. Once salt water has penetrated an underground area, it becomes exceedingly difficult to purify the water. "The continued inland advance of the ocean water into the western basin," reads a late report, "will result in the ultimate destruction of the fresh water

supply." The west basin is a rapidly expanding industrial area
and includes three large Douglas Aircraft plants, the North American Aircraft plant, most of the oil refineries and rubber plants,
as well as such large establishments as Columbia Steel, Johns-
Mansville Products, General Chemical Company, Joshua Hendy
Iron Works, and many similar plants. Unless the intrusion of salt
water is arrested, this whole area will, in a few years time, become
an island undermined by salt water.

A section in Ventura County, known as the Oxnard Claim, is
today seriously threatened by saline intrusion brought about by
excessive pumping. Here the water level is already from 10 to 20
feet below sea level and will continue to drop if pumping continues at the present rate. The neutralization of alkaline soils by
the use of chemicals is an extremely expensive process and requires
a great deal of time. The deeper wells are drilled, in many areas,
the greater becomes the danger of irrigation waters being contami-
nated with a heavy boron content. There is a strong possibility that
the Oxnard Claim, for example, which was originally reclaimed
from the sea, may revert to the sea. The improvident location of
industrial plants and the lack of effective regulation have brought
about a situation in which chemical wastes have drained into underground water basins, ruining the water supply. Recently Dr. Fritz
W. Went, of the California Institute of Technology, announced
that Colorado River water may contain substances injurious to
plants and shrubs brought about by artificial water softening processes. Garden plants such as ferns, begonias, azaleas, and larger
shrubs like the cherry laurel, are known to be injured by sodium
created when calcium and magnesium are transformed in the
water-softening process. Various clovers used in lawns seem, also,
to be affected. Dr. Went believes that the spread of Bermuda
grass may be brought about by the fact that other grasses, such as
blue grama and fescue, are adversely affected by artificially sof-
tened water. The balance of ecological forces, in an environment
like California's, is extremely delicate and, in many respects, has
been upset by an improvident use of water. Notoriously diligent
in the quest for water, California is guilty of an enormous waste
of water. In Los Angeles alone, over 400 cubic feet per second of

waste waters are discharged into the ocean despite the fact that studies have shown that a large part of this water could be reclaimed for industrial uses.

Just as more water is required during "dry cycles" for irrigation, so the urban per capita consumption also increases. The per capita consumption in Los Angeles increased from 140 gallons a day in a three-months period in 1946 to 156 gallons a day for the same period in 1947. The drier the season, the more water is needed. Also as a city expands and industry increases, the demand for water constantly exceeds the amount which would be required by the mere increase in population; the urbanization of an area has an important effect on the water supply. For example, it has been estimated that each 1,000 feet of safe yield water will supply the domestic and industrial requirements of about 7,000 people or water for 700 acres of land. Or, to put it another way, about the same amount of water is required to irrigate an acre of agricultural land that is required for an acre of city land in use. The choice, therefore, between agricultural or industrial use is clear cut. Many communities are faced, therefore, with a difficult decision involving the use of water. Considerations of this kind account for the "water strategy" of Los Angeles where two-thirds of the water supply is imported from outside the area. The strategy hinges on the use of Owens Valley water for domestic and industrial uses; Colorado River water for supplemental domestic and industrial use; and underground waters for irrigation and as a source of supply for outlying towns and communities. Since the underground supply can be tapped at cheaper rates than must be charged for imported water, the policy has always been to earmark the underground supply, as far as possible, for irrigation purposes.

How Not to Conserve Water

Since California has never had a state policy governing the use of underground waters, the kinds of crops planted and the acreages devoted to these crops have been largely determined by the market prices for these crops. Large sections of the state's best lands have been planted to crops requiring a ruinous amount of water simply because these happened to be the "best" cash crops. With the government committed to a policy of purchasing "surplus" potatoes in 1948, California growers planted potatoes in a crazily improvident manner and, for the same reason, the cotton acreage soared to new heights. In Kern County, hills were leveled, foothill pasture districts were invaded, and hundreds of new wells were drilled and new pumps installed to produce an ever-greater "surplus" of "surplus" potatoes. Relatively unfertile lands were "spiked" and "hopped up" with soil foods and fertilizers to increase the tonnage of a crop that every grower knew would exceed all rational demands. It is scarcely necessary, therefore, to emphasize the effect of government price policies on water uses. Both potatoes and cotton are heavy water-consuming crops in California: it takes 30 inches of water in the growing season to make a cotton crop. Yet, in years of a water famine, cotton production in California for 1947 was 50 per cent greater than 1946 and the 1948 production topped that of 1947 by a similar margin. Not only did this increase place an enormous drain on water resources, but, as a result of improvident planting, sections of the San Joaquin Valley have already begun "to blow," and a minor dustbowl is in the making.[2]

The amount of water which can be properly utilized in the production of various crops is determined, of course, by a number of variable factors: the kind of crop; the price of irrigation water; the nature of the soil; the type of plowing and cultivation; and so forth. There has never been a systematic attempt in California to plan production in reference to these variables so as to conserve the maximum amount of water. Furthermore, water has been used, in many cases, in an utterly wasteful and often harmful manner. For example, lettuce can only use 15 inches of rain or the equivalent to advantage; sugar beets, 28 inches; clover, 25 inches; cotton, 30 inches. Although agricultural experts have spread this information far and wide, there are lettuce growers who have been known to pump the equivalent of 73 inches of rainfall, enough water, as Stanton Delaplane has pointed out, "to grow rice and go swimming besides." Current proposals in California to impose a

[2] *Ibid.*, December 30, 1948.

tax on those who waste water suggest how important the planning of production is in relation to a real water conservation program.[3] In the absence of such a program, of course, the "big squeeze," as Mr. Delaplane writes, "is on the land which is being wrung dry for every drop of water as the farmer exchanges his underground water bank to the level of his commercial account in the bank; the squeeze to plow and irrigate more and more acres of grazing land and to feed and help employ a state doubled in population. This is the squeeze that squeezes everybody."

THE OUTLYING COASTAL DISTRICTS

What a drouth can do to urban communities in California can best be illustrated by a brief account of what happened in Santa Barbara, Ventura, and San Diego in 1948, all coastal communities. Located on the Pacific, north of Los Angeles, Santa Barbara and Ventura occupy a particularly disadvantageous position in the water scheme of California. Neither area can benefit from Colorado River water, for the distance is too great. The stream flow is negligible and there is little underground flow in either area. Most of the available water comes from rainfall and the stream flow from the south slopes of the mountains.

Between July 1, 1947, and March 14, 1948, only 2.36 inches of rain fell in the barren watershed of the Santa Ynez River, sole source of Santa Barbara's water supply (by comparison with a "normal" rainfall of 18 inches). By March the river had literally dried up. Gibraltar Dam, twenty miles east of Santa Barbara, at one time held 14,000 acre-feet of water; but, by 1948, the reservoir was so badly silted that the storage capacity had been reduced to half that amount. Furthermore the water had become sulphurous, unpleasant to smell, and capable of coloring almost anything it touched. Homes near the dam's canal were coated black by the fumes. Three supplemental wells were soon pumped dry and the city drilled six new wells, but still the supply was inadequate. Farmers in the area, fearing that the city would drill further

[3] A few counties already have ordinances prohibiting the waste of water: *see:* Los Angeles *Times*, November 28, 1948.

wells, took action to protect the underground supplies and, also, to prevent the sale of water to Santa Barbara residents.

Even earlier, on January 16, 1948, Santa Barbara adopted an emergency ordinance rationing the supply of water. Famous for its lawns and gardens and luxury hotels, Santa Barbara was compelled to limit the use of water to household and sanitary purposes, and to impose a $300 penalty for the use of water to irrigate lawns. With the grass on the plush golf courses turning a sickly yellow, wealthy estate owners trucked water in oil tankers at two cents a gallon to keep their favorite plants and shrubs alive. Specific uses of water, such as for washing sidewalks, driveways, and automobiles, were forbidden. Despite these measures water consumption soared above the pre-ordinance rates. The ordinance aimed at keeping the daily consumption at 1,500,000 gallons, but the actual consumption was double this amount. Nevertheless green lawns began to alternate with brown to give Santa Barbara a crazy-quilt pattern.

By March, bath water was being siphoned into garden hoses and either fed to thirsty plants or stored in old tubs and barrels. Sink drains were disconnected so that the run-off might be caught in pans for later re-use. Owners of $40 camellia bushes began to scour the countryside for water. Volunteer compliance committees were organized to check water meters and to ferret out bootleggers. Street sweepers discontinued sprinkling and restaurants ceased to wash down their tile floors. By March 6th, the city was described as being "so dry a blade of green grass looks like an oasis." With the spread of rationing, the city's water revenues dropped and water rates had to be doubled to meet operating expenses. Water wells drilled to a depth of 300-feet were abandoned when the drillers failed to strike water-bearing rocks. The use of water for shower baths was restricted, and two-inch tub-baths or "teasers" became the order of the day. An incalculable investment in lawns and gardens and nursery establishments was threatened; water for oil drilling was cut off for the duration of the emergency; and city officials visited farm areas trying to "borrow" water.

In the midst of this emergency, a $2,000,000 bond issue was submitted to the voters for the purpose of raising additional revenue to finance new water supplies. The proposal was soundly defeated by the voters! Trying to find out why the voters had rejected the bond issue, a newspaper reporter discovered that Santa Barbara is full of water magicians. One man interviewed favored sticking a pipe into the mountainside and simply sucking out the water. Another was convinced that a secret lake existed beneath Gibraltar Dam. Still a third claimed to hold a patent on a device which would take air in on one side and turn out water on the other. Still others interviewed were convinced that the solution was to be found in taking the salt out of ocean water. Despite the fact that the Chamber of Commerce, concerned about the million-dollar tourist trade, imported camels to parade through the streets with signs reading "Even a camel needs a drink," the bond issue failed.

The Anglo-American lawn-sprinklers and water-wasters who have invaded Southern California lack the inherited social sense of the importance of water that the Spanish and Mexicans have always possessed. A visitor to Santa Barbara in 1945, when the drouth really began, would never have suspected that he was gazing upon an arid or semi-arid environment. Wherever he looked, he would have seen beautiful gardens, flowers, plants, shrubs, and, in Montecito, he could have visited some of the most magnificent estates in America. Yet the millionaires in Montecito's famous "butler belt" were harder hit than the residents of Santa Barbara for they were restricted to 50 gallons per person per day (the Santa Barbarans were allotted 140 gallons). The vote on the bond issue in Santa Barbara shows clearly enough that the "water problem" in Southern California is in part sociological and cultural: people do not understand the importance of water. It would be most unfair, however, to censure these people, for the scarcity of water in most Southern California communities is a closely-guarded top-level secret.

In nearby Ventura, water-rationing was also adopted as an emergency measure. By March the city's reservoirs were down to 8,000,-000 gallons—the minimum amount which the city must maintain for fire-fighting purposes. In a race against time, the city bored still another well to a depth of 1,422 feet and fortunately tapped a fairly good underground flow. But the well was 7,400 feet from the nearest city main and the officials did not have pipe to make the connection. At one time, the city had only enough water to last for eight days. One section of the city went completely "dry" for several days and the public school in the area had to be closed.

Paradoxically the six principal communities in Ventura County are located in a flood control zone. It was in this area that the St. Francis Dam collapsed on March 12, 1928, releasing a torrent of water that destroyed 1,240 homes and took 385 lives: one of the worst flood disasters in California history. The county has three "spreading grounds" in the area in which surface waters are stored underground. Eighty per cent of the county's agricultural production is located in this flood control zone, including some of the most profitable citrus enterprises in California. Yet this flood area was short 30,000 acre-feet of water on July 23rd, 1948 and the county's hydraulic engineer announced that the citrus belt was directly imperiled by the water shortage. The fact that millions of dollars have been spent in this area for flood control serves to indicate that there is an adequate water supply if it were properly conserved and properly utilized; but the "catch" in this proposition is overall integrated planning and watershed control.

Similarly there is a solution for Santa Barbara's critical water problem. The Bureau of Reclamation has plans for the construction of a $30,000,000 project on the Santa Ynez river 36 miles from Santa Barbara. From a dam at this site, the Bureau proposes to bore a six-mile tunnel through the mountains to bring 210,000 acre-feet of water a year to the city of Santa Barbara, a supply of 10 million gallons a day. But the city would have to agree to purchase the water and to bear a portion of the cost (principally the costs of aqueducts and conduits) estimated at $15,000,000. To date the project is still "pending."

San Diego is still another community faced with a critical water problem. Until quite recently, San Diego was dependent upon 10 reservoirs which stored surface and run-off waters, for its water supply. In 1940 the city's population was around 200,000 and it had projected certain long-range plans for the development of a future water supply. But, during the war, the population jumped to 400,000. Naval installations alone consumed 40 per cent of the water supply and new industries placed an added burden on the city's water system. By May, 1948, El Capitan Reservoir, the largest in the city's system, contained only 5 per cent of its normal storage, and the Sweetwater Reservoir was so low that farmers were raising melons on the floor of the dam. With the Navy threatening to abandon San Diego as a base, the city rushed to completion a 71-mile aqueduct to tap the Colorado River water which Los Angeles was bringing to the coast. Today this aqueduct brings some 65 million gallons of water a day to San Diego, but this is not enough to fill the smallest reservoir and, at the same time, supply the daily needs of the population. Already the city is planning to build a second aqueduct to bring in a larger supply of Colorado River water.

But these are not the only areas of critical water shortage in Southern California. On August 7, 1948, officials of the La Cañada and Flintridge foothill areas in Los Angeles met with the Board of Supervisors to work out a plan whereby the growth of these areas might be arrested! With 23,000 water-users in the area, the officials explained that an additional two or three thousand residents would create a drouth. To check population increase, the planners proposed a tightening-up of building and zoning ordinances. This is probably the only case on record where a Southern California community has sought to arrest its growth by artificial means. In the Mar Vista area of Los Angeles, residents were without water for a 12-hour period on August 31, 1948. Youngsters went scurrying throughout the neighborhood filling pans and pails with water and housewives had no water for cooking. In West Riverside, on September 5th, the water supply failed for a 30-hour period and an emergency "bucket brigade" was organized to bring water to the homes. In one area after another, city officials warned that the shortage of water had created a fire peril for supplies were too low to permit the extinguishing of a single major fire.

Drouth conditions set in motion a chain-reaction sequence in California. Water supplies are taxed far beyond normal demands. The drier it gets, the more water must be used. When the rains fail to come on schedule, the hazard of frost increases. The longer the "dry spell," the greater the danger of forest fires. The deeper wells are drilled, the less efficient pumping becomes (that is, the more water is lost) and the greater the cost. The more power is used for pumping, the greater becomes the strain on the utility systems whose storage capacities are endangered by precisely the same conditions which produce the drouth. The more water that is dumped over dams to turn generators, the lower becomes the emergency water supply. This is why water is not merely a "problem" in California but a code-word to designate a hundred problems.

The greater the water crisis, the more difficult it becomes to do anything about the problem, for shortage intensifies the social cleavages. There are sharp conflicts between water-users in California: between urban and rural areas; large cities and small towns; between different types of water-users; between watershed and non-watershed users; between those who rely on surface waters and those who use underground waters. One community's plans will stimulate fear and apprehension in the next community. All communities try to build up vested water rights, and, since the right depends upon use, they have been known to create artificial uses to bolster claims they intend to make in the future. Whatever one community does about water is almost certain to concern some other community. In building its aqueduct to the Colorado, Los Angeles was forced to bore a tunnel in the San Jacinto Mountains. Now Riverside contends that the tunnel drains 200,000 acre-feet of water to which it has a prior claim. These internecine rivalries, suspicions, quarrels and plots severely limit the state's capacity to plan for the needs of all, and this, despite the fact that the needs are obviously interrelated and interdependent.

THE FLOOD PARADOXICAL

Just as California has always had "dry cycles" so it has always had floods. In fact, the floods have caused almost as much damage as the dry spells. The floods of California are, of course, exceptional. California's floods do not have their counterpart in any other section of the country. In other areas, floods are of long duration, and are characterized by a spread of exceedingly high waters over a great expanse of territory. But California floods are usually of short duration, of high water velocity, and are concentrated in a few areas. Along the coast, elevations rise from sea level to eight and ten thousand feet within 25 or 30 miles. The run-off from these mountainous slopes can be devastating both in volume and velocity. In other areas, warnings of approaching floods can usually be given some days in advance; but floods develop in California in a matter of hours. Floods also recede in California almost as rapidly as they appear. Because of the steep mountain slopes, flood waters in Southern California carry an enormous amount of debris: rocks, giant boulders, trees, brush, rubbish, and dirt. By improvidently contoured roads in the foot-hill and mountain areas, the run-off of flood waters has been greatly accelerated. Miles of paved streets and improved land in the pathway of floods have also had the same effect. In the absence of proper planning, large communities have been permitted to come into existence along the sides and at the mouths of canyons. Huge industrial plants, in some cases, have been built directly in the path of flood waters. It has only been, moreover, within the last few years that the state has developed a comprehensive flood control program.

Talcum-dry under the summer sun, the two major rivers of Los Angeles, the San Gabriel and the Los Angeles, wind like giant arteries from the mountains to the sea. These are, beyond all doubt, the most highly improbable rivers in America, rivers whose "fish need feet." Dry for most of the year, they can carry an enormous volume of flood waters. The highest record of rainfall for a period of 24 hours' intensity in the United States was recorded one year in the Santa Anita canyon near Pasadena, when 26 inches of rain fell in a 24-hour period. By 1941, over $260,000,000 had been spent in harnessing the seasonal flow of these two rivers, perhaps the "driest" rivers in America. More than a billion dollars' worth of land lies within the overflow path of the two rivers and nearly a million people live within the 300,000 acre potential flood zone. The San Gabriel is only about 39 miles long, from Mt. San Antonio to the point where it empties into the Pacific near Long Beach. But more money has been spent to control its flood waters than have been spent on rivers four times its length and carrying ten times its volume of water.

Where dams have been constructed to impound flood waters, it often happens that the flood waters bring down to the reservoir tons of leaves, rocks covered with silt, debris, and various types of organic matter that decompose rapidly, forming various products of deterioration, and making the water unfit for consumption. In time these products of decomposition spread to the underground areas near the dam and thereby contaminate the wells. Poorly planned flood control projects have added materially to the cost of water-treatment processes. Furthermore the paving of flood channels tends to reduce the natural ground percolation which is vitally important in building up underground water storage basins. Owners of valuable homes and businesses along water channels scream for better and bigger flood control projects; they are little concerned with saving water. Powerful vested interests have also been created to support crazily planned flood control projects, for flood control means cement, rock, sand, gravel, and other supplies. For years, now, Los Angeles has been carrying an ever-larger volume of flood waters to the Pacific in cement-lined conduits, canals, and drainage systems, waters that it vitally needs to conserve, waters that might be diverted into underground storage basins.

The direct damage caused by floods in Southern California—the driest area of the state—has been enormous. A major flood in Los Angeles County in 1934 took 30 lives and caused property damage running into the millions. In a five-day period in March, 1938, 11 inches of rain fell in Los Angeles, flooding 30,000 square miles, taking a toll of 81 lives, and causing an estimated $83,000,000 in property damage. The indirect damage caused by flood is also enormous: the loss of water; the spoilation of impounded waters in storage reservoirs; the damage to the watershed. The control of floods, like the utilization of water, demands integrated regional planning and controls.

BRUSH FIRES AND BURNS

In 1945 more than 9,500 forest, brush, and range fires burned over some 600,000 acres in California, causing direct property losses in timber, forage, watershed cover, and farm crops running into the millions of dollars. In 1945 it cost $2,000,000 to put out the fires in California's national forests alone. The danger from fires is, of course, in direct ratio to the dryness of the region and the season. Southern California is particularly exposed to the hazards of brush and forest fires. On May 17, 1945, 388 brush and grass fires were reported burning in Los Angeles County. In May, and again in late August, when the red-hot desert winds sweep in through Cajon Pass, the brush and forest fires invariably rage.

Both floods and forest fires in Southern California have increased in direct relation to the increase in population. In the Angeles, Cleveland, Los Padres, and San Bernardino National Forests, all in Southern California, 252 fires in 1947 burned over 13,593 acres, causing known property damage of $334,663 to the watershed and destroying homes valued at $500,000. In September, 1948, forest fires destroyed 50 or more homes in Ojai and caused extensive property damage. In July, 1947, a brush fire burned over 10 square miles in the Chatsworth area and destroyed 50 or more homes. Throughout Southern California, some 3,300 fire-fighters were employed in July, August, and September of 1948. A raging fire in Big Tujunga Canyon, in August, 1947, sent great billowing clouds of smoke over Los Angeles, burned over 3,500 acres, cost two lives, and injured 208 people.

Throughout Southern California one can see the great scars or "burns" caused by forest fires: the Beatty Burn, the Bryant Burn, and many others. The Forest Service reports that it will take more than 15 years to repair the damage which these burns have caused to the watershed. The drier the season, of course, the greater the danger of fires; the more likelihood there is of fires, the more severe become the restrictions placed on travel in forest areas. Already 50 per cent of the Angeles National Forest has been forbidden to vacationists and, should the drouth continue, the whole area may be closed to tourists and visitors. Southern Californians seem to find it difficult to believe that the "brush" which covers the steep slopes of the mountains is as valuable as a heavy stand of timber. "This chaparral," reports the Chief Forester of the area, "is just as important to us as big trees. You can't build houses with it. But it protects us from floods. Burn it, and we're heading for trouble. Our steep mountain cover takes a good 15 years to regrow after a major fire." It goes without saying, of course, that the most intimate relation prevails between fires and floods and the scarcity of water.

Despite the menace of forest and brush fires in Southern California, where fires of this character have caused more damage than in any area of the state, administrative functions in relation to fire prevention are divided between city, county, state, and federal agencies in the craziest fashion. A full-time coordinator was not employed until as late as 1948. The prevention of forest fires is still not integrated with the control of floods and the conservation of water. Foresters complain bitterly of the catastrophic procedures being followed by flood control engineers with their "tin-roof" theory of the disposal of flood waters but each service continues to function in sublime independence of the other. The problem is not one of extinguishing but of preventing forest fires; more accurately, it is a problem of integrated regional control of the entire watershed in its broadest possible implications. Watersheds are irreplaceable assets and their neglect in California, notably in Southern California, is something to make the angels weep.

This chapter has been designed to demonstrate the nature of the "water problem" in California. Other western states have a water problem, but in most of these states the problem is of miniature proportions by comparison with California. In many of these states the problem is essentially one of developing latent water resources but the problem in California has implications that reach into every phase of the economic life of the state. Here the problem has psychological, sociological, and cultural implications which are no less clearly defined than the engineering and technological phases. California has pioneered in the *development* of water resources, and has long had a comprehensive state-wide water plan. But it has not planned, on a similarly comprehensive scale, for the proper utilization and conservation of a limited water supply. Once this type of planning is undertaken, it will be quickly discovered that the planning of water resources has unsuspected dimensions in California.

[16]

THE POLITICS OF THE COLORADO

<><><><><><><><><><><><><><><><><><><><><><><><><><><><><><><><><><><><>

*AND a river went out of Eden
to water that garden.*
—GENESIS 2:10

THERE IS a startling paradox about the social consequences of the struggle for water in an arid environment. As a youngster in northwestern Colorado, I often heard my elders recounting tales of violent quarrels between old friends and neighbors over the perennial question of "water rights." Somehow the phrase, "water rights," so innocuous in itself, seemed to carry overtones of blood and violence. In an arid environment, men will fight for water with a truly implacable bitterness, a bitterness beyond reason and entreaty. For if there is not enough water to meet all needs, there is really no basis for compromise: there is nothing to negotiate. Water controversies, therefore, present the ultimate in the way of irreconcilable points of view. On the other hand, nothing will weld disparate elements into a more cohesive force than a common concern over water. If men will fight over water, they will also cooperate to conserve it and the history of water controversies is that, in the long run, the rule of cooperation prevails. In an arid environment, water is the ultimate sovereign. The embittered 27-year-old controversy between Arizona and California over the water of the Colorado River provides, perhaps, the classic illustration of this paradox.

MENACE OR RESOURCE?

Dominating land and man, the Colorado River is the greatest single fact within an area of nearly a quarter million square miles.
—FRANK WATERS

Just as men will fight to the death or cooperate untiringly over water, so, in such an environment, water itself has the paradoxical quality of being either a menace or a resource. This is notably true of the Colorado, an ugly, tricky, and defiant river, yet at the same time, an efficient and tireless worker. The Colorado, as Frank Waters has written, is "bigger than its statistics," for it is at once "an international headache, a geographic skeleton, a hydrographic puzzle, a roll call of the most familiar names in the whole Southwest, and a symphony complete from the tiniest high pizzicato of snow-water strings to the tremendous bass of thunderous cataracts reverberating in deep canons." Nowadays people speak of "the law of the river" by which they mean that the Colorado, in the last analysis, is a law unto itself, sovereign over water "rights," compacts, legislation, and court decisions.

With its sources high in the snow-capped Rockies, the Colorado empties into the Gulf of California below the Mexican border, 1,400 miles from its source; the second longest river in the United States. The river drains a vast area of 244,000 square miles: one-twelfth of the land area of the nation, an area greater than Spain and Portugal combined. The basin of the Colorado is some 900 miles long and varies in width from about 300 miles in the upper to about 500 miles in the lower basin. Actually it drains not one but many basins for it is made up of some 50 rivers or tributary streams extending into seven western states: Arizona, California, Colorado, Nevada, New Mexico, Utah, and Wyoming, and it is also an international stream. The total population of the basin is still below a million people, with an average density of fewer than four persons per square mile and with only two cities of more than 20,000 residents. The basin area, in fact, is one of the most sparsely settled regions in North America.

Within the basin area, one can find almost every extreme of altitude, temperature, precipitation, geological formation, scenery, and wild life. Temperatures vary from 50 degrees below zero to 130 degrees above; the growing season ranges from 80 to 300 days in the year; and, in the course of its 1,400 mile journey, the river drops a good two and a half miles from above 14,000 feet to .248 feet below sea level. The Colorado is the "driest" river system in America. The annual precipitation for the entire basin is less than 15 inches—the lowest of any of the river systems—and 90 per cent of this precipitation is lost through evaporation. The remaining 10 per cent of the scant precipitation makes up the vital flow of the Colorado. The violent extremes which characterize everything connected with this remarkable river are reflected in the great fluctuations in the volume of water it carries: the annual flow at Lee's Ferry varies from 5,500,000 acre-feet to 25,000,000 acre-feet. Thus when one speaks of the "average" or "normal" flow of the Colorado, the words are purely figurative: there is nothing normal or average about the river and this, of course, is a cause of controversy. Note, also, that seven states have an interest in the Colorado, as well as the federal government and the Republic of Mexico.

Seven arid, water-hungry states have a vital interest in the Colorado, but one of these states, California, has a most peculiar relation to the river, or more accurately, *one section* of the state for the Colorado is of almost exclusive concern to Southern California. Although the Colorado is the last water resource of Southern California, the peculiar fact is that, with the exception of a small 4,000 square mile section on the desert, California is entirely outside the watershed and contributes scarcely a drop of water to the river. On the other hand, almost the entire area of Arizona is within the watershed of the river and this area makes up 45 per cent of the basin drained by the Colorado. Water from the Colorado used in Southern California cannot return to the river. Even Imperial Valley, which Colorado River water has converted from desert to truck garden, lies outside the watershed of the river. Water flowing into the Imperial Valley by the All-American Canal cannot return by gravity flow to the parent stream but, by a freak of nature, drains northward and empties into the Salton Sea. The unique relation of Southern California to the Colorado is of immense political and psychological importance, first, in the sense that the other western states have always regarded California as an interloper or outsider; and, second, because California, being an outsider, has always stressed its interest in the Colorado rather than the development of the region itself. In fact it is difficult for

Southern Californians to think in terms of the welfare of the Colorado Basin because they are strangers to this basin; hence, they can only see their interest in the river.

To understand the background of the Colorado River imbroglio, however, it is necessary to keep a further fact in mind: although California is an outsider, it has been largely responsible for the development which has taken place on the river up to the present time. Without governmental or other subsidy, California has invested over half a billion dollars in the development of the river and initiated the first projects for the control and utilization of the river as a great natural resource. If there had not been a powerful center of population in Southern California, clamoring for water and able to provide an unlimited market for power, it is altogether likely that Boulder Dam would still be in the blueprint stage. The Metropolitan Water District of Southern California not only entered into a contract with the federal government to purchase the entire output of power of the dam at a good price, but provided the funds for the construction of the necessary transmission lines and appurtenant works, including the generative equipment in the dam itself. In fact, this equipment is operated by the district under contract with the federal government.

At the turn of the century, only 261,197 people were living in the Colorado River Basin: one person per square mile. In political terms, it would have been asking the impossible for the six western states to have requested Congress to appropriate the huge funds necessary for the construction of Boulder Dam in terms of the well-being of this handful of people. In congressional thinking, "economic feasibility" is an important factor in all projects of this kind. Southern California, the one large center of population in the Southwest, alone possessed a resource-base, which in terms of the taxing power, made Boulder Dam a practical and feasible undertaking. Boulder Dam was the *first* great multiple-purpose project of its kind in this country. It is clearly the parent of TVA and all similar developments. Today the feasibility of huge development projects of this kind has been demonstrated but in the 1920's, with Calvin Coolidge in the White House, Congress had to be shown. Today one reads the debates on the Boulder Dam bill with utter amazement: it seems incredible that Congress could have been so skeptical, so mulish, so penny-wise, pound-foolish; but this was the political reality of the times. It is only in terms of this reality, moreover, that the vital role which Southern California played in the development of the Colorado can be understood. And this role, incidentally, provides a key to an understanding of Southern California's possessive attitude about the Colorado.

WATER IMPERIALISM

"The City of Angels," writes Leverett G. Richards, "wears its water as ostentatiously as a newly-rich widow displays her jewels. Cafés flaunt waterfalls cascading down their fronts. Water flows lavishly in fountains and in Hollywood swimming pools. Obsessed with rain, which seldom falls outdoors, Los Angelenos boast the world's only indoor rain—showers every ten minutes on the minute. The Angels do everything with water, in fact, but drink it. There is no rationing of water, no conservation, no restriction." With this lavish and prodigal use of a scarce resource, how does it happen that Southern California is desperate for water? Why is it necessary for this region to stand guard twenty-four hours of the day over the water resources it has appropriated? How, in other words, does one account for the water imperialism of Southern California?

Free-flowing artesian wells changed the whole economic outlook of Southern California in the decade just prior to the turn of the century. By 1900 some 11,000 artesian wells were still flowing, and it was these wells which had made the desert that is Southern California blossom like the rose. With local wells pumping 78,-000,000 gallons of water, Los Angeles had a water supply in 1900 adequate for a population of 500,000 people (the population was then around 150,000); but the city had already begun to think about future needs. Some of the tycoons of the city, however, were also thinking about the nearby San Fernando Valley. So, in the early years of the century, Los Angeles made its first water raid, its first imperialist conquest. At an initial cost of $25,000,000 (the ultimate cost was to be $90,000,000), Los Angeles built a

238-mile aqueduct to Owens Valley, on the eastern slope of the Sierras, and brought 288,000,000 gallons of water a day into Southern California.

There can be no doubt that this raid was an act of imperialism for not only was Owens Valley far removed from the watershed of Los Angeles but the project was brought off by fraud and violence and ruined a once-prosperous farming community. For many years Owens Valley water was used, not to supply domestic needs in Los Angeles, but to irrigate farm lands in the San Fernando Valley, to the $50,000,000 enrichment of the plotters and schemers who had gobbled up thousands of acres of unimproved land. Moreover, to induce the residents of Los Angeles to bond themselves to the tune of $25,000,000, an artificial water famine had been created. In fact, the residents of Los Angeles were innocent parties in this scheme, since they were kept as much in the dark about what was happening, and why, as were the farmers in Owens Valley. It is important to note the "spring" or mechanism of this plot: the use of an artificially created urban water famine as a cover for a trans-mountain raid on water put to use, during the interval when population expectation had not been realized, to irrigate farm lands in Southern California at the expense of farming communities in the raided area.

Owens Valley water made possible the spectacular growth of population and industry in Los Angeles in the period from 1900 to 1920. But by 1913, when the project was fully completed, Los Angeles, then a city of 350,000, was already dreaming of a future population of 2,000,000. The next raid consisted in the building of another 150-mile stretch of tunnels, siphons, and canals which tapped the waters of Mono Lake, a hundred miles beyond Owens Valley, high up on the eastern slope of the Sierras. Then in 1923 the city created the Metropolitan Water District, approved $266,000,000 in bonds, and built a 410-mile aqueduct to the Colorado. This aqueduct, the longest in the world, tunnels mountains, crosses earthquake faults, and traverses miles of desert terrain. By the time the aqueduct was placed in operation in 1941, virtually all of the artesian wells had long since ceased to flow and 30,000 pumps were tapping the underground water supply. In some areas, as in Pasadena, the wells were down to 350-feet and the water table, generally, was falling throughout the area. Built to supply a supplemental water service to 2,350 square miles in the Southern California littoral, the aqueduct can carry approximately 1 billion gallons of water a day—1,212,000 acre-feet a year —enough to support a population of 4,500,000.

At the present time the Los Angeles Basin area has combined water sources which, if fully utilized, would support a population of 10,000,000: an estimated gross supply of about 1.4 acre-feet of water per acre of land. Nowadays the total Owens River supply, estimated at 284,000,000 gallons per day, is consumed in Los Angeles. And, since the Owens Valley aqueduct was built, a vast underground network of run-off streams has been tapped and brought under control through an elaborate system of perforated pipes and pumping wells. In addition some 37 reservoirs and 44 storage tanks have been built along the line of the Owens Valley aqueduct, capable of storing 131,000,000 gallons of water and, within the city limits of Los Angeles, are reservoirs which store a month's reserve supply of water. Currently the city consumes about 131,000,000 gallons a day: an average daily consumption of 140 gallons per customer. The Owens Valley supply alone makes up 73.6 per cent of the city's current water needs.

In effect, therefore, Los Angeles has available water resources which can be estimated somewhat as follows: Owens Valley supplies adequate to support a population of 2,000,000; underground water supplies to support a population of 350,000; and Colorado River sources, if confirmed, adequate to support a population of 4,500,000. As will be noted, Los Angeles is getting from 80 to 90 per cent of its current supply from sources other than the Colorado River. Its concern with Colorado River water, therefore, is, like its concern with Owens Valley water in 1900, a concern for the future, not for the present. And concerned for the future it should be, with people pouring into Los Angeles at the rate of 16,000 a month. This constant preoccupation with *future needs* has, indeed, characterized Southern California's water strategy—its water imperialism—since the turn of the century.

Now Arizona is concerned not about the future but about immediate *necessities*. Only 1 and 2/10ths per cent of the total area of the state is under irrigation. Central Arizona, with 725,000 acres under irrigation, makes up 81 per cent of the total irrigated acreage in the state. Originally desert, this section of Arizona is the heart and core of the state's agricultural economy. Here a heavy use of available waters has resulted in a rapidly falling water table and an overdraft on the underground sources which is estimated at 468,000 acre-feet a year. The amount of water stored in all the reservoirs of the Salt River Valley has dropped from 1,560,000 acre-feet in 1941 to 393,899 feet in 1946. With only eight inches of annual rainfall, it requires 4 acre-feet of water per acre per year to irrigate these lands. The extensive re-use of pumped waters concentrates the salt content and now threatens thousands of acres of valuable farm land unless new sources can be obtained to flush out the salt. Additional thousands of acres will certainly go out of production unless new water sources are developed. To remedy this situation, the Bureau of Reclamation has designed the Central Arizona Project, with an estimated cost of $738,000,000, which would divert water from the Colorado River, pump it to an elevation of 985 feet and carry it by gravity-flow through a 235-mile canal to Central Arizona.

Naturally Arizona wants this project *now*, but Southern California is bitterly opposed to it, now or in the future. In opposing this and similar Arizona projects, Southern California stresses the needs of domestic water users as a paramount right. The inference, of course, is that Southern California is not using Colorado River water for irrigation purposes. The "catch" in the original Owens Valley raid, it will be recalled, was the existence of potentially valuable farm lands in San Fernando Valley. Is there a similar "catch" in Southern California's latest campaign for water? Residents of Southern California, constantly inundated with propaganda from such organizations as the Colorado River Association, are never told anything about the relation of Imperial Valley to the water strategy of the association. In fact, the Southern Californians are as much in the dark about Imperial Valley as the residents of 1900 were about the rich farm lands of San Fernando Valley. Just what part, therefore, does Imperial Valley—"The

Palm in the Hand of God"—play in the water imperialism of Southern California?

NATURE'S FREAK: THE IMPERIAL VALLEY

Located in the extreme southeastern part of California, somewhat larger in area than the State of Delaware, the Imperial Valley is one of nature's strangest freaks. The valley is about 45 miles long and 30 miles wide and occupies the southern portion of a large below-sea-level desert basin extending from the shores of the Salton Sea to the border. Slightly above the floor of the valley lie two desert plains known as the East Mesa and the West Mesa. The Salton Sea, at the northern end of the valley, originally came into being when the Colorado River went on one of its periodic rampages and flooded the valley. Although the Imperial Valley appears to be level, it actually slopes northward toward the Salton Sea at a rate of from 4 to 10 feet per mile. The difference in elevation, from south to north, is about 200 feet, making the valley ideally suited to irrigated farming. The entire valley drains into the Salton Sea which, of course, has no outlet. With the exception of Calexico on the border, most of the towns of the valley are below sea level. The average yearly rainfall is only about three inches. Walled off from the coast by a high mountain range, Imperial Valley is a desert pit or oven, a kind of natural hothouse and nursery. Once a desert, the valley now produces crops worth $100,-000,000 annually, and is one of the most valuable winter truck gardens in America. The crop acreage is about 424,202 acres but, since two and three crops a year are produced, the actual acreage is nearly double what it appears to be. Imperial Valley, it should be noted, occupies a marginal or peripheral relation to Southern California and, for that matter, to the rest of California: it is really a part of the Colorado River basin.

Basic to an understanding of Imperial Valley is the fact that its development, the first large-scale development on the Colorado, was undertaken under *private* auspices. In one sense, Imperial Valley was the *first* large-scale development project in the west, for its development dates from 1896, and the Reclamation Act was not passed until 1902. The development of the valley was a most heroic undertaking, and the story of this development is fascinating and dramatic.[1] But the point to be noted here is simply that this gigantic development scheme should never have been undertaken as a private project: the scale was too vast, the commitments too immense, the problems too enormous. There was no "easy" money involved in the financing of this project, and no benevolent federal government to provide assistance. As a consequence, the project was bogged down from the very outset in financial intrigue, endless litigation, foreclosures, private feuds of rival factions, and similar difficulties. Bold as the original scheme was, it soon proved to be inadequate, and the development company was forced to cut corners, to improvise, and to take fatal risks. Then, in the winter of 1905–1906, the Colorado broke through its channel 4 miles below the international border and for 16 months poured its full flow into the Imperial Valley causing untold havoc and damage. The development company had to sell its holdings to the Southern Pacific Company which in turn sold these holdings to the Imperial Valley Irrigation District formed in 1911—the largest irrigation district in the world. Ironically, the disastrous flood of 1905–1906 proved to be a godsend, for it was this flood which really set in motion the agitation that finally resulted in the passage of the Boulder Dam bill in 1928.

Misbegotten at the outset, Imperial Valley has always had the character of an aborted community; a half-formed, twisted, ill-conceived mongrel. The social affairs of the valley have always been as badly snarled as its financial affairs. Indeed, the tangle of social relations reflects the original cross-purposes which came of the fateful attempt to undertake for private profit a development which was essentially public in purpose. Of the farm owners, 40 per cent do not live in the valley, yet they control over half the acreage in farms. By comparison with more stable farming communities, the degree of absentee ownership is abnormally high. In 1936 about 43 per cent of the farm land was owned by slightly less than six per cent of the total number of farm owners. Corporate

[1] *See:* David O. Woodbury's excellent volume, *The Colorado Conquest,* 1941; and *The Colorado* (1946) by Frank Waters.

farms, also, control about 23 per cent of the cropped acreage. No one lives in the valley in the summer, of course, who can afford to escape to the coast. The summers are long, hot, and dry, with temperatures ranging from 100 degrees Fahrenheit to as high as 120 degrees. The agriculture of the valley is highly specialized and speculative, making for a most unstable type of farming. The winters are short and mild, with killing frosts seldom occurring between November 15th and March 15th. In fact, frosts only occur during a period of 10 days or less throughout the year, giving the valley a growing season of about 300 days: perhaps the longest growing season of any farming community in the nation.

This extraordinary valley was transformed in twenty years from a barren desert into a fabulously rich winter garden. On January 1st, 1901, only a few Indians were to be found living in the entire valley; but by 1910 the population was 12,000; by 1920, 40,000. In 1901, there was not a single planted acre in the valley; by 1920, 410,000 acres were in production. The entire valley, in its every aspect, is completely dependent upon Colorado River water. The rapid forced-growth of the valley is an important key to an understanding of the social maladjustments, the abnormal land tenure pattern, and the pronounced cleavages, tensions, and antagonisms which have prevailed for so many years. Today the population is about 47 per cent "native white"; 35 per cent Mexican; with a mixture of Japanese, Negro, Indian, Filipino, Hindu, and other elements. Today as yesterday, the valley is ruled by a set of power-drunk ruthless nabobs who exploit farm labor with the same savagery that they exploit the natural resources of the valley.

The Imperial Valley, it will be noted, is in precisely the same position as the central portion of Arizona: both sections are desert lands which have been converted into truck gardens by the magic of Colorado River water. Both sections raise essentially the same kind of crops, by the same methods, and for the same markets. There is this all-important difference, however, that the development of the Imperial Valley was undertaken much earlier than the central portion of Arizona, and Imperial Valley taps the Colorado River below the point from which central Arizona proposes to take water. Imperial Valley, since it gets "the last crack" at Colorado River water this side of the border, is in an excellent position to benefit from any "surplus" in the stream, assuming there is a surplus.

It must also be emphasized that Arizona is on the *wrong* side of the Colorado River to make the maximum and cheapest use of river water. For Arizona is a tableland that slopes toward the Colorado, so that pumping operations are required to lift water to elevations from which it can flow to the areas where it is needed. Imperial Valley, on the other hand, is below sea level and, although it is outside the watershed, it is much easier to divert Colorado River water into the valley than into Arizona lands. It was for this reason, in fact, that Imperial Valley's development was the first on the river.

As the pioneers in water development in the Southwest, the Imperial Valley growers know the importance of water, and forty years of fighting for their "rights" has schooled them in the arts and tricks of water warfare. They are determined, resourceful, and ruthless, and they have the advantage of a headstart in development which has created, in their favor, certain vested rights which legislatures as well as courts are reluctant to disturb. They insisted that the All-American Canal should be made an integral part of the Boulder Dam project. They were also insistent that the canal should be built to capacity size. Eighty miles in length, the canal is concrete-lined, over 20 feet deep, and 200 feet wide at the water's surface. Nearly 3,000 miles of lesser canals and laterals carry water throughout the valley. Because of the freakish nature of the valley, the water would run much too fast in this canal if it were not controlled, so five "drops" have been built along the line of the canal which generate about 4,000,000 kilowatt-hours of electricity per year. The Imperial Valley Irrigation District, which owned the old International Canal, is obligated to repay the cost of the All-American Canal which amounts to some $36,000,000.

The lion's share of Colorado River water used in California is currently consumed, not in Los Angeles, but in Imperial Valley; not for domestic but for irrigation purposes. Of recent years, Imperial Valley has used as much as 2,717,530 acre-feet of water annually by comparison with only 56,000 acre-feet being diverted by the Metropolitan Water District. In fact, the Los Angeles aqueduct, of recent years, has taken only two per cent of the city's allotted maximum supply from the Colorado. It is this huge diversion of water to Imperial Valley for agricultural purposes which is never mentioned in Southern California's water propaganda. For if this diversion were emphasized, then the issue would be whether California has a paramount right to develop its arid lands at the expense of similar lands in Arizona. Not only does Imperial Valley use an inordinate volume of water, but it actually wastes and misuses water. Colorado River water used in Imperial Valley cannot find its way back to the river but drains into the Salton Sea. In 1947 it was estimated that 1,116,000 acre-feet of water drained, in this manner, into the Salton Sea. This, if it could be re-used, would be of enormous value. Furthermore the ground water table in many areas of the valley is only from 3 to 4 feet below the surface so that as much as 80,000 acres are waterlogged and have been taken out of production. For years the valley has been troubled with alkali in the soil and great patches of salt-ridden alkali wastes can be seen in the valley today. But the absentee owners are not concerned about the long-range effect of over-irrigation; they are only concerned with immediate profits.

But this is not all: Imperial Valley now proposes to develop lands on its East and West Mesas, totalling some 500,000 acres and it would like—in fact, it intends—to irrigate these desert lands with additional water from the Colorado. But, of some 225,000 acres on the East Mesa, the soil surveys of the Bureau of Reclamation indicate that only 35,900 acres are of a character that warrants reclamation. The West Mesa report is not yet available but it will probably indicate that only a small portion of the 200,000 acres in this area should be placed under irrigation. When these soil surveys are mentioned in Southern California, the propagandists scream "bureaucracy," "government crack-pots," and so forth, and seek to make it appear that the Bureau of Reclamation is arbitrarily retarding the development of the valley. The feud between the valley and the Bureau dates back, in fact, to the time when President Theodore Roosevelt unsuccessfully tried to wrest control of the irrigation development from private hands and to vest it in the Bureau of Reclamation. The Imperial Valley nabobs know perfectly well that the 500,000 acres on the East and West Mesas are of doubtful long-range productivity, but they are equally well aware of the speculative profits to be made in peddling this land. Their primary concern, however, is to use these lands to build up a claim to an established, vested use in additional water from the Colorado.

Further north from Imperial Valley is the Coachella Valley where a remarkable development has taken place in the last seven years. The Coachella Valley is an entity in itself: a small, pocket-like valley that produces 90 per cent of the dates produced in the Western Hemisphere. In 1946 the date crop was worth about $4,000,000. Dates, of course, are enormous water-consumers. There is an old Arab saying that a date palm must have "its feet in water and its head in the fires of heaven." One acre of date palms requires from 10 to 20 feet of water. With such an enormous drain on the water supply, it is not surprising that the underground water table began to drop at an alarming rate some years ago. To remedy this situation, a new 74-mile canal has been constructed which takes off from the All-American Canal about 20 miles west of Yuma, and brings water to the Coachella Valley by gravity flow. The valley is ideally adapted to irrigation, with perfect drainage, low humidity, and high temperatures. Some 20,000 acres are now in production but, with the first Colorado River water arriving through the canal on March 20, 1949, a 100,000 additional acres will soon be brought into production. In this instance, at least, a nearly 95 per cent use will be made of the water, for the canals and laterals are cement-lined, underground perforated pipes have been installed to collect the seepage, and spray-irrigation is in wide use throughout the valley. The valley produces, in addition to dates, many other crops with an aggregate value (1948) of $13,500,000. Unirrigated lands in the valley sell for $75 an acre but prices for the date acreage range from $3,000 to $5,000 an acre. From the point of view of value per acre, these are the richest agricultural lands in the United States, producing an annual income of $1,000 and more per acre. A study of land ownership in the valley, made by the Bureau of Reclamation, indicates that there are 696 holdings of 40 acres or less making up a total of 14,437 acres; but there are 4 holdings which in the aggregate total 21,041 acres. This will give some idea of who is benefiting from Colorado River water in the Coachella Valley. The use of Colorado River water in the Coachella Valley is, of course, on an exact par with the intended use of this water in Central Arizona.

How does it happen, therefore, that urban industrial Los Angeles works hand-in-hand with the Imperial and Coachella Valley interests in the fight for Colorado River water? Theoretically these two different uses would seem to be in conflict, and they may well be in actual conflict in the future. One of the basic reasons for the alliance is that the rights of Los Angeles, however they may be defined, are *junior to* the rights of Imperial Valley. Another explanation is to be found in the fact that some of the most powerful behind-the-scenes forces in Los Angeles have important interests in the Imperial Valley. It should also be kept in mind that "vested rights" play an important role in water disputes. To justify a claim for water, it is important to demonstrate a "need" for water and, under the doctrine of appropriation, a present *use*. Imperial Valley's present uses buttress urban Los Angeles' long-range requirements. To build up these claims, Imperial Valley has been permitted and encouraged to waste water for the same reason that an artificial water famine was created in Los Angeles in 1901 to bolster the city's claims to Owens Valley water. In 1945, there were 18,556 fewer acres under irrigation in Imperial Valley than in 1946; yet Imperial Valley managed to "use" more water in 1947 than 1946. To be exact, there was an additional use of 446,980 acre-feet of water. California claims a total of 5,362,000 acre-feet of Colorado River water but of this total all but 1,112,000 acre-feet is claimed by the Imperial, Coachella, Palo Verde, and Yuma districts for agricultural purposes. This is the real "catch" in the California position.

But the question still remains: why does Los Angeles encourage this water-waster, the Imperial Valley, when, in the long run, the interests of the city may conflict with those of the valley? The answer to this question involves another: who benefits in Los Angeles from Colorado River water? The City of Los Angeles classifies water-users as residential, commercial, and industrial. Beyond all doubt, the industrial concerns are the great water-consumers. It takes something like 77,000 gallons of water to refine 100 barrels of oil. The plant of the Union Oil Company in Long Beach consumes as much water as the entire City of Long Beach. Among the heavy industrial users are the oil companies, the rubber companies, and the steel mills. It is, therefore, significant that Mr. Joseph Jensen, chairman of the board of the Metropolitan Water District, should also be the Chief Petroleum Engineer for the Tidewater Associated Oil Company. Residential consumers are taxed for water in Los Angeles in a four-fold manner. Their tax bill includes an assessment for the Metropolitan Water District; their ordinary city taxes include items which represent water costs; then they pay their water bills; and, finally, their power bills. The total cost of the Colorado River Aqueduct up to 1941 was $189,704,000. Tax assessments levied by the Metropolitan Water District amounted to $124,000,000 by November 30, 1948, a large portion of which represented interest on the bonded indebtedness. In the year 1947, the aqueduct operated at 14 per cent of its installed capacity, or 5.53 per cent of full capacity, so that the city actually sustained a loss of $8,783,807.93 on the importation of Colorado River water for that year. The major part of this burden falls, not on the industrial users, but on the general taxpayers of the community. To keep the taxpayers quiescent, it is necessary to maintain an incessant ballyhoo about the "water crisis" and, for this reason, Los Angeles is willing to use Imperial Valley as the means by which it can build up facts and figures indicating the existence of such a crisis.

THE OVER-DIVISION OF WATER

One could write a tome of several thousand pages on the legal and economic phases of the Colorado River water controversy alone. The pertinent documents now make up a good-sized library. I do not intend to attempt such an analysis here, but it is extremely important that certain basic facts about this controversy should be kept in mind. Paradoxical as the statement may sound, the Colorado River controversy really turns on the conflicting "rights" of various states to non-existent water. For the fact is that, at the present time, there is simply not enough water in the river to meet the demands of all parties. The controversy is, therefore, rather like the embittered fights which often develop between the partners of a mine over the profits from a mine which has not yet been placed in production. As the history of the West abundantly shows, there are few more bitter conflicts than those which men will wage over hypothetical or purely speculative bonanzas. The fact is, as clearly stated in the Bureau of Reclamation's great report [2] that "there is not enough water available in the Colorado River system for full expansion of existing and authorized projects and for all potential projects outlined in the report, including the new possibilities for exporting water to adjacent watersheds." Yet, and here is a further paradox, more than 8,000,000 acre-feet of water now flows unused every year across the international boundary.

Under the Colorado River Compact, which made possible the development of Boulder Dam, the Colorado was divided into an Upper and Lower Basin. Lee's Ferry was selected as the point of division since, at this point, all of the "upper" tributaries have entered the flow of the Colorado. The compact did *not* apportion water between states but merely between *basins*. By the terms of the compact, the upper basin states (Utah, Colorado, New Mexico, Wyoming), were allotted 7,500,000 acre-feet a year. The lower basin, consisting of California, Nevada, and Arizona, being given 8,500,000 acre-feet. At the time this compact was made, the law on interstate waters was more or less in its infancy, and the records on the flow of the river which were then available were incomplete and, in some respects, inaccurate. As since determined, the average annual flow of the river at Lee's Ferry is considerably less than the negotiators assumed. Instead of apportioning water by percentages, which has since become the standard practice, the negotiators made the mistake of dealing in acre-feet, a mistake which

[2] *The Colorado River* (March, 1946), p. 21.

creates the illusion that a certain amount of water is bound to be in the river whether it is or not and thereby makes any compromise difficult to effect.

The Colorado River Compact was ratified by six of the states involved but Arizona refused to ratify the company until 1944. In order to secure the passage of the Boulder Dam Act, California was forced to adopt what is known as the California Self-Limitation Act by the terms of which California limited its claims on the Colorado River to 4,400,000 acre-feet of the 8,500,000 acre-feet apportioned to the Lower Basin states plus one-half of any "surplus" water that might be available. The compact negotiators apportioned water between the basins on the assumption that, once this division had been made, the states could reach an agreement on their respective shares. But no agreement has been reached by the lower basin states; hence the controversy.

To complicate matters, the United States negotiated a treaty with Mexico in 1944 by the terms of which Mexico was allocated 1,500,000 acre-feet per year from the flow of the Colorado. It was the negotiation of this treaty, more than anything else, which drove Arizona into the compact, for it feared the effect, on its rights, of this additional burden on the river. California, needless to say, strenuously opposed the Mexican Treaty and appropriated $75,000 to conduct a lobbying and publicity campaign against ratification—the first time that a state has actively opposed ratification of a treaty. At the time, California was receiving nearly one-third of all the waters of the Colorado, to which she contributes not a drop, and she was not yet using more than half of that third. The Imperial Valley Irrigation District actually offered to supply Mexican needs out of its own facilities, but at a price of approximately $340,000 a year. Moreover, as Frank Waters has pointed out, half or more of all the water allocated to Mexico is "return flow," that is, water that has already been put to use within California and has seeped back into the river and canals on the way to Mexico. The Mexican allocation was certainly not excessive for it involved only 3 or 4 per cent of the primary flow of the river. But, as usual, there was a "catch" in California's position: the Imperial Valley Irrigation District owns, through a subsidiary, the old International Canal in Mexico. It wants, therefore, to have full control over the distribution of *any* water which Mexico receives, not only to redeem its investment in the old canal and to be able to make a service charge on the delivery of water, but also to obtain the maximum flow of water through its main canal so as to generate some 33,000 kilowatts of power annually at Pilot Knob.

Since the Bureau of Reclamation has estimated that the total long-time average flow of the Colorado at the U. S.-Mexican border is 17,200,000 acre-feet per annum, it follows that the waters of the river have now been allocated as follows: 7,500,000 acre-feet to the Upper Basin states; 8,500,000 acre-feet to the Lower Basin states; and 1,500,000 acre-feet to Mexico, making a total of 17,500,000 acre-feet and leaving a "surplus" of only 220,000 acre-feet. California now holds contracts with the Bureau of Reclamation, negotiated when the Californian, Ray Lyman Wilbur, was Secretary of the Department of Interior, for the delivery of 5,362,000 acre-feet of water from the Colorado—a figure considerably in excess of the 4,400,000 acre-feet stipulated in the California Self-Limitation Act. How does this difference arise? In controversy between Arizona and California is approximately 2,000,000 acre-feet of water: 1,275,000 acre-feet being involved in the "consumptive use" versus "depletion" controversy; about 500,-000 acre-feet involved in the controversy over the so-called III-B "water" of the compact; and something over 600,000 acre-feet involved in the controversy over the loss of waters through evaporation.

These three issues make up the bone of contention between Arizona and California. To understand the first of these issues, the "consumptive use" versus "depletion" controversy, and the controversy over III-B water, it is necessary to note certain facts about the Gila River. The Gila is a tributary of the Colorado but it arises in New Mexico, flows through Arizona, and then finally empties into the Colorado *below a point at which any state* other than Arizona could possibly make use of its waters. It was from the flow of this stream that Arizona had developed its Central Arizona farming area. At the time the Colorado River Compact was negotiated, the waters of the Gila had been fully *appropriated* and were then *in use*. Arizona, quite naturally, insisted that the waters of the Gila belonged to it *because the Gila River was Arizona's*, and the facts of geography made it impossible for any other state to use the waters of this river. To satisfy Arizona on this point, the Colorado River Compact negotiators, in apportioning water to the Lower Basin, gave to the states of this basin, under Article III-A of the compact, "the exclusive beneficial use" of 7,500,000 acre-feet (the same apportionment given the Upper Basin states); and then provided in Article III-B that the Lower Basin should have "the right to increase its beneficial consumptive use of such waters by one million acre-feet per annum." This additional 1,000,000 acre-feet—the so-called III-B water—was apportioned the Lower Basin to compensate Arizona for the flow of the Gila which, at the point where it empties into the Colorado, was estimated to be, in a virgin state, approximately of this volume. The III-B provision was actually *added to* the compact to satisfy Arizona.

Why, then, does the provision allocate this water to the "Lower Basin" and not specifically to Arizona? For the reason that the division of waters made by the compact was *between basins* and it was, therefore, feared that if any waters were apportioned to a specific state, by name, it would touch off a general controversy. The Arizona negotiators however, were given the unanimous assurance of the other negotiators that the additional million acre-feet was for Arizona alone. It was also planned that, immediately following the signing of the compact, this understanding would be embodied in an agreement between California, Nevada, and Arizona, dividing the waters apportioned the Lower Basin. This supplemental agreement was never negotiated, however, and, therefore, Arizona refused to ratify the compact. It is now California's contention that this III-B water is not water apportioned to the Lower Basin but falls within the category of "surplus" water of which it is entitled to receive one-half or 500,000 acre-feet.

As to this phase of the controversy, the facts clearly substantiate Arizona's contention. It would have been an act of unprecedented generosity on Arizona's part to have consented to the inclusion of the Gila within the Colorado River system for purposes of apportionment since, not only are the waters of the Gila of no concern to any other state, but these waters were fully appropriated and in use at the time. Furthermore, when the governors of the seven states met in Denver, on August 20, 1927, in an effort to arbitrate the controversy preparatory to the signing of the Boulder Dam Act, they approved Arizona's contention; and, in adopting the Boulder Dam Act (December 21, 1928), Congress wrote into this act a specific provision (sec. 4-a) requiring California to adopt a Self-Limitation Act by which it would limit its demands to 4,400,-000 acre-feet "of the waters apportioned to the lower basin states by paragraph *a* of Article III of the Colorado River Compact, plus not more than one-half of any excess or surplus waters unapportioned by said compact." This provision was inserted for the express benefit of the other states and to secure their support for the Boulder Dam Act by allaying their fears concerning California's intentions. The adoption by California of a self-limitation act *in these terms* was made an express condition of the Boulder Dam Act. It is most significant, therefore, that the Boulder Dam Act limited California to III-A waters and carried into effect, with minor changes, the recommendations of the governors' conference. And the report of the governors' conference was quite specific on the point that the million acre-feet of Paragraph III-B had reference to the flow of the Gila. It is also significant that the other western states, with the exception of Nevada, which is California's satellite, have consistently upheld Arizona's contention.

It will be recalled that, in apportioning water between the basins, the compact uses the phrase "beneficial consumptive use" which, although its meaning is crucial, is not defined in the compact. In its controversy with Arizona, California contends that this phrase has reference to the aggregate of all the individual items of consumptive use *at the various points* of diversion or use. If California is right about the III-B waters, therefore, it would charge Arizona with 2,375,000 acre-feet of water, which is the total use which Arizona now makes of the waters of the Gila, and deduct this amount from Arizona's share of the waters of the Colorado. The Gila River is what is known as a "wasting" river for much of its flow is lost through evaporation and seepage along the sandy reaches of the lower part of the stream. It is agreed that the virgin flow of the Gila, before any improvements were made, was about 1,275,000 acre-feet at the point where the Gila empties into the Colorado. But Arizona engineers, through an elaborate network of irrigation systems, canals, and pumping operations, have greatly increased the consumptive use, in many areas, by use and re-use of the same water. Arizona contends, therefore, that "consumptive use" refers to the *depletion* of the river, not at the point of each separate diversion, but at its mouth, and only by those causes which have to do with man's activities. The Arizona contention is based on the sound proposition that, in an arid environment, *salvaged* waters should not be charged against the state which has developed their use. Why, it will be asked, is California such an artful bookkeeper when it comes to adding up the uses made of water on the tributaries of the Colorado? The answer is quite simple and also quite conclusive: *there are no tributaries of the Colorado in California.*

There remains the question of how the states should share in the loss of Colorado River water, stored in on-stream and off-stream reservoirs, occasioned by evaporation. The compact makes no reference to this important question, a question which involves between 600,000 and 800,000 acre-feet of water a year, for the evaporation rate is quite high. California has the interesting view that it should not share in any losses occasioned by evaporation when water is stored in reservoirs beyond its borders, although the water is being stored there for its benefit as well as for the benefit of the other states. Arizona contends that the various states should share in these losses in the ratio that they receive water from the Colorado. On this point it is very difficult to see how it makes any difference whether water is held in storage in off-stream or on-stream reservoirs; stored water is stored water. The mere fact that California diverts water *below* Boulder Dam can hardly justify the contention that it should not share in losses occasioned by evaporation at Lake Mead, the huge storage reservoir behind the dam.

There is, of course, an air of grotesque unreality about this controversy. In the first place, the basic causes of the controversy relate back to calculations made at the time of the compact concerning the "average," "normal," long-term flow of the river, and estimates of "surpluses." These terms have very little meaning when applied to the dynamic Colorado. In the second place, the Colorado River Compact is a reflection of what might be called "historical lag." When the compact was negotiated in 1922, the conception of a river valley authority hardly existed. What the Colorado River really demands is watershed control in all its various and interrelated aspects. The mechanism of an interstate compact is entirely too static, too legalistic, too inflexible to cope with a river like the Colorado. What is needed is a river authority or some similar mechanism which could focus the attention of the parties, not upon their "rights," but upon the problems of the river.

While Arizona and California are wrangling over their rights, the Colorado is piling up silt behind Boulder Dam at the fantastic rate of 400,000 tons a day. The Colorado has perhaps the highest silt-content of any river in the world and carried, prior to the construction of Boulder Dam, an estimated 100,000 acre-feet of silt annually. Many of the tributaries of the river have highly eroded watersheds. Thus, Boulder Dam has a limited life-expectancy. As the silt piles up, additional up-stream dams must be built to protect Boulder Dam, and, incidentally, to protect the Lower Basin from what could be the most disastrous flood of all time. But this is not all: over grazing on the Navajo Reservation is a prime cause of the silting in Lake Mead. To cope with the problem of silt, controls would have to be extended over watershed uses of the land. Since Boulder Dam was built, the water below the dam carries less silt than formerly and hence flows more rapidly and its cutting-power is much greater. Already the reports indicate that the Colorado is beginning to cut back its channel at the point where it empties into the Gulf of California. This cutting-back process will accelerate unless it is controlled. In short, the Colorado is sovereign in this controversy; it will not respect "rights" or "compacts" or court decisions. A Colorado River Authority could integrate all aspects of the problem of controlling the river and its tributaries and its watershed. Such an authority could shift

[315]

water from one area to another; from one use to another; as needs changed and circumstances demanded. But there is wholly lacking, in the compact, the flexibility which is required for control of the river.

The essential vice in California's position in the Colorado River controversy, apart from the speciousness of some of the arguments it has advanced, is that it has never really admitted that the Colorado River belongs to the Southwest, as a region, and that it is not the exclusive property of one state. California has never thought that it was part of *any* region; hence its "isolationist" attitude, its water imperialism. One can sympathize with California's attitude and for many reasons: it does occupy a *peripheral* relation to the other states, being a more or less self-contained unit or region in itself; it *must* necessarily think in terms of *this* region, not some other region; and it has unquestionably shown great courage, vision, and daring in initiating developments which have redounded to the benefit, not only of other states, but of the nation. California's fight for the Boulder Dam Project blazed the way for TVA, for Grand Coulee, and for all the great multiple-purpose projects which are now pending. And, quite apart from these factors, there is the abiding reality that the Colorado is Southern California's last water hole. With new residents pouring into Los Angeles at the rate of 16,000 a month, who can blame Southern California for being concerned about its *future* water supply? Large cities must necessarily project water developments far into the future for it takes time to develop these projects; it cannot be done overnight. But, as a loyal Southern Californian, I do deplore the casuistry, the lack of candor, the occasional double-dealing, and, above all, the reliance upon brute power, which have all too frequently characterized the region's quest for water.

THE CENTRAL VALLEY PROJECT

◇◇◇

THE IMPORTANCE of Colorado River water to Southern California is matched by the importance of the Central Valley Project to northern and central California. The development of the Colorado and completion of the Central Valley Project involve vastly important and enormously complex issues: social, fiscal, political, and technological. But vital as both projects are to California, there is little overlapping or connection between them; each stands, so to speak, on its own legs. Little of the water in the Colorado originates in California; but all of the water involved in the Central Valley Project has its origin within the state. There is a sense, however, in which both projects are intimately related, for they present different aspects of the same problem, namely, California's relations with the federal government and with the other western states. Antagonisms formed by California in the Colorado River fight could have serious repercussions on the Central Valley Project. The point to be noted in both cases is that California seeks to function as an independent, sovereign empire. The possessiveness that it has shown about the Central Valley Project finds its counterpart in the arrogance of the state's attitude toward the Colorado. The Colorado River controversy presents the outward aspect of California's relations with the federal government and the other western states; the Central Valley Project involves the internal dynamics of this relationship.

CALIFORNIA—UPSIDE DOWN

In terms of Colonel J. W. Powell's famous definition of an arid region—a region with less than 20 inches of annual rainfall—California is really an arid state. But the aridity of the state shades off as one moves from north to south. The need for irrigation, of minor importance in the north, becomes absolute as one moves south. In the areas of scantiest rainfall, the rate of evaporation is greatest; hence a larger supply of water is needed. Although it takes less water to irrigate in the northern part of the state, the bulk of the irrigable land is in the southern part. California, therefore, is a state that is upside down, a state in which nature seems to be at cross-purposes with man. There are at least two senses, however, in which this contrariness of nature is a major asset. First, because it presents a constant challenge to man's inventiveness and thereby stimulates technological advancement; and, second, because there are inherent advantages in this seemingly crazy disposition of resources. For example, the long dry season is an enormous agricultural asset. It forces growth, makes possible sun-drying or outdoor hothouses. Similarly, the levelness of the land, in certain sections, is a great advantage although it presents serious drainage problems. But there is more to this apparent contrariness of nature than first meets the eye.

In a famous article which appeared in the Census of 1884, E. W. Hilgard pointed out the great advantages of irrigated agriculture. Irrigation, he said, makes for much greater fertility; more intensive production; smaller units of operation, thereby pyramiding the number of people that the land will support; more reliable production; and, also, it enables the farmer, as he put it, "to impart to the penny a nimbleness unheard of in regions dependent upon the seasons alone." Irrigation actually multiplies the available farm acreage in an arid environment by making it possible to raise two and three crops from the land in a single year. The advantages of irrigation are, by reason of the nature of man, seldom fully realized except in areas where *irrigation is an absolute necessity.* Where the necessity is partial, man will be inclined to gamble on rain; it is only where the challenge is absolute that the full advantages of irrigation can be realized. California, of course, is the classic illustration of this principle.

The upside-down character of California is epitomized in the great Central Valley which, being a freak of nature, presents a powerful incentive to plan. The Central Valley is a great oblong bowl, 500 miles long and approximately a hundred miles wide, enclosed by the Sierra Nevada mountains on the east, the Coast Range on the west, the Klamath Mountains to the north, and the Tehachapi Range to the south. It produces about 50 per cent of the state's cash farm income, covers 18 counties, contains 83 cities and towns, and supports about 1,500,000 Californians. With an average growing season of 250 days (by comparison with 140 days in the East and Middle West), it produced crops in 1947 valued at $916,689,948; only four states in the union exceeded this production. The valley has about 50 per cent of the water resources of the state, and the streams which enter the valley from the east contain nearly all of California's potential hydroelectric power. By any test, the valley is one of the most fertile valleys in the world and it is, beyond all doubt, the "heartland" of California. It might best be described as a great air-conditioned, outdoor hothouse.[1]

Although Californians refer to this region as *The* Central Valley, it is really two valleys: the Sacramento and the San Joaquin. The singular is generally used for the reason that the southern valley, the San Joaquin, is "hidden" or concealed; it actually appears to be part of the Sacramento Valley. A low alluvial divide formed by the delta of the Kings River, not apparent to the eye, separates the drainage of the southern part of the valley from the northern. The Sacramento River flows from north to south; the San Joaquin from south to north. Both streams meet opposite San Francisco and find their way to the ocean through a break in the Coast Range, the Golden Gate. Although some of the streams of this vast area find their way out of the valley, others have no means of escape for the valley, a great "longitudinal trough," extends a hundred miles *south* of the area drained by the San Joaquin River. The lowest points in this southern part of the valley were originally occupied by swamps, sloughs, and "dry lakes" or evaporating pans, which received the flow of streams having no exit from the valley. Most of the valley is treeless and large portions are almost completely level. The northern portion has about

[1] *See:* "The Land Where Whoppers Come True" by Camerson Shipp, *Saturday Evening Post,* Nov. 20, 1948.

one-third of the land and two-thirds of the water; the southern part, two-thirds of the land and about one-third of the water. Both valleys combined have about 63 per cent of the irrigable land of the state and 51.6 per cent of the state's water resources, so that, even if the water resources were fully developed, there would still be a slight deficit of water. However, this deficiency is much greater in the San Joaquin, the lower Valley, which has 16.8 per cent of the water resources and 36.3 per cent of the irrigable land. The lower valley must make up this deficit by getting water from some other watershed. Fortunately the flow of the San Joaquin River is so slight, that it can be reversed by lifting water a total of only 160 feet in 160 miles. The Central Valley Project is designed to pump water from the Sacramento up the San Joaquin Valley to a point a distance of 110 miles, where the water will be released into the channel of the San Joaquin through which it will run into San Francisco Bay. This pumped water will replace the natural flow of the San Joaquin which will be diverted southward through a 160-mile canal to irrigate lands in the southern part of the valley. Thus, by an ingenious piece of engineering surgery, water will be shifted southward in two stages: from the Sacramento to the San Joaquin; from the San Joaquin to the lower valley.

GENESIS OF THE PROJECT

The story of the struggle to harness the water and power resources of the Central Valley is one of the most dramatic chapters in the history of western America. Modern irrigation really began in the valley in 1871 with the completion of the San Joaquin and Kings River Canals, in the Fresno area. In large part, this development was made possible by the "colony" type of settlement in which groups or "colonies" of settlers were established in particular areas. The colony form of settlement, by comparison with the isolated farmstead, made for a remarkable sense of community interest in the development of an adequate water supply, and it also made possible the organization of the necessary manpower—two essentials of pioneer irrigation developments. Once the canal was built, plans were developed to extend it, and a survey made by

The Central Valley Project

Lt. Col. B. S. Alexander in 1874 set forth, in all essentials, the original conception of what is now the Central Valley Project. It will be noted that this scheme was projected thirty years before the passage of the Reclamation Act (1902), when the federal government had not as yet developed a reclamation policy.

The second important phase in the development of the project took place in 1878 when William Hammond Hall, the first State Engineer, induced the legislature to appropriate $100,000 to make the first comprehensive water plan for the valley. Hall, a man of real technical ability and rare social insight, insisted from the outset on the importance of an overall, integrated plan and emphasized the dangers of a haphazard, piece-meal development. But Hall's maps and notes for the plan were never published, largely because he had courageously insisted upon a revision of the water laws of the state as a necessary pre-condition to the plan. The Southern Pacific Company, which then owned thousands of acres of land in the valley and held a virtual monopoly on riparian water rights in many areas, objected to any revision of the water laws and, since the company controlled the legislature, it was able to secure the abolition of the office of State Engineer.

For many years following the abolition of the office of State Engineer, every attempt to develop a water plan for the valley was consistently and successfully obstructed by various special interest groups, although the idea of such a plan was never forgotten. During this period, the water and power resources of the valley were exploited for private profit rather than for the greatest measure of public interest. Private power companies secured rights to use water for power-generation on most of the tributaries of the Sacramento and San Joaquin Rivers, and, also, staked out claims on the Kern River in the southern end of the valley. At the same time, the large landowning interests bottled up the available water resources. As Hall had foreseen, the development of water for irrigation became bogged down in an embittered controversy over water rights. This controversy had to do, of course, with the conflict between the doctrine of "riparian" rights, which the state had adopted as part of the common law, and the doctrine of appropriation. Not only did the prolongation of this controversy tie up land titles and arrest the development of water resources, but the courts eventually upheld the doctrine of riparian rights, and thereby set back the whole idea of a comprehensive water plan. Certain private rights, which became "vested" at this time, have not yet been extinguished. Thus the fact that the first major water projects were undertaken by private interests, and for profit, definitely retarded and still complicates the development of the valley.

Although a great agricultural development took place in the valley following 1878, it became apparent by 1920 that further development was impossible within the existing framework of laws and institutions. By 1920 the private-power interests had gobbled up most of the available power sites; the mining interests had ruined thousands of acres of orchard and wheatlands by uncontrolled hydraulic mining. These mining operations, as well as ill-conceived drainage projects, had largely destroyed the navigable value of the Sacramento and San Joaquin Rivers; all of the comparatively cheap water had been appropriated; large landowning interests were still obstructing power, water, and drainage projects; and the further development of the valley, as a consequence of the shameful period of private gouging, had been brought to a temporary halt. As so often happens in California, realization of the impasse which had come into being was accentuated by the occurrence, in the twenties, of a cycle of "dry years."

During this period of almost unrestricted private exploitation, however, the groundwork had been laid for a further and larger development. In 1887 the legislature had passed an act authorizing the formation of irrigation districts, and investing these districts with important powers (the pioneer legislation of this kind in western America). Henry Miller, the great land baron, bitterly opposed the formation of irrigation districts, and took a test case, involving the constitutionality of the districts, to the United States Supreme Court. Irrigation, he contended, belonged of right to "private enterprise"; but the Supreme Court ruled otherwise. Under this legislation, small farmers could pool their resources and accomplish in their own behalf projects of the type that Henry Miller had undertaken for his personal enrichment. With an average of 60,000 acres being added each year to the irrigated areas, the formation of irrigation districts eased the tension but, once the cheap-water resources had been exhausted, the districts lacked the financial power to undertake needed developments. In the period from 1900 to 1921, also, an immense amount of indispensable data was accumulated on the use of water for irrigation, on the average stream-flow of virtually every stream in the state, and on elevations. A Debris Commission, formed in 1893, had also undertaken pioneer work on the improvement of the channels of the two main rivers. Similarly the formation of drainage districts, making possible various drainage and reclamation projects, had added materially to the acres in cultivation. Furthermore, a great new interest had been aroused in water-and-power developments as shown by the creation of a Conservation Commission in 1911 and a State Water Commission in 1913 and the calling of a State Water Problems Conference in 1916. It should be noted, however, that the act creating the state water commission was held up by a referendum, initiated by private interests, but was finally ratified and approved by the voters in November, 1914. During these years, also, the Supreme Court of the state was forced to modify, in important respects, its earlier decisions upholding the doctrine of riparian rights.

Although developments of this character were important in themselves and also laid the foundation for later developments, it was apparent by 1920 that these piece-meal unrelated developments created new problems without really solving old ones. It took the people a long time to realize that *any* single project, whether it involved power development, flood control, drainage, irrigation, or the improvement of the channel of a river, automatically disturbed the balance between every other factor involved. A drainage project, planned as a drainage project, might solve one problem but it would also aggravate many related problems. The piece-meal development of the period from 1878 to 1920 was all right as far as it went, but it had reached the limits of expansion. It also became apparent that, unlike similar schemes elsewhere, the development of the Central Valley could not be undertaken as a "watershed" project since several different watersheds were involved. What was needed, therefore, was a plan for the development of the entire valley. With this realization, a special, "peculiar," dilemma arose.

Elsewhere in the West, areas in need of large-scale development projects had automatically turned to the Bureau of Reclamation for assistance. As a matter of fact, California interests had submitted a plan to the Bureau in 1905 but, since this plan called for the expenditure of an estimated $40,000,000, and the entire funds then available to the Bureau for the 11 western states was only $28,000,000, the Bureau had been compelled to reject the proposal. As has happened on more than one occasion, the *scale* of things in California could not be adjusted to the national program for the other western states. Furthermore, a very *special* problem existed in relation to the development of the Central Valley because of its purely *intrastate* aspect. This was a purely one-state project, and the basis of federal aid was, and still is, premised on narrow constitutional considerations, namely, flood control and navigation, two of the least significant aspects of the project. As a federal project, the Central Valley Project also presented a special political problem for the state was by no means unified in support of the project. Southern California, which gets its water from the *eastern* slope of the Sierras and from the Colorado River, was not at all interested in the Central Valley Project. Even the Central Valley failed to present a unified support for the project since a nexus of conflicting interests, public and private, could not agree on the nature of the development. Faced with this complex of interests, the federal agencies naturally hesitated to intervene. More than anything else, however, it was the very magnitude of the project which discouraged the Bureau of Reclamation.

Faced with this impasse, it is deeply significant that California should have undertaken, in its own right, initially unaided by the federal government, a project of the magnitude of the Central Valley Project. But the need, the challenge, was so great that California could not hesitate, and the internal dynamics of expansion were of such a character that events dictated decisions. Long before the federal government was actively interested in the project, the state had spent over $1,000,000 merely in preliminary investigations and surveys. As in the case of Boulder Dam, the imperial manner in which California asserted its demands reflected the urgency of its needs and, since both projects, Boulder Dam and the Central Valley Project, originated in California, it is easy to understand why the state has always insisted, contrary to the facts, that they are California projects. As it worked out, the fact that the Central Valley Project was purely intrastate in character, actually enabled California to seize the initiative and to push forward a project which, if it had had to depend upon the consent of other states, might have been held up for many years.

THE INTREPID COLONEL

If Lt. Col. B. S. Alexander was the first to conceive the outline of the Central Valley Project, Colonel Robert Bradford Marshall was the man who first aroused widespread popular support for the project. In the autumn of 1891, Colonel Marshall was sent to California as Chief Geographer for the U. S. Geological Survey on a map-making project that carried him into every corner of the state. Setting out for Stockton from Nevada City in a double buckboard wagon drawn by four mules, he camped the first night on the high bluffs near Folsom and got, the next morning, his first view of the great Central Valley. Marshall was a hydrological expert, not a project engineer, but he had been deeply impressed with the work of Major J. W. Powell, the Director of the U. S. Geological Survey, who believed that the only way to win the West was by irrigation. Wherever Marshall went, in his surveys, he kept records of land levels and water measurements, and studied irrigation projects. In Southern California, where he was stationed for a time, he had occasion to study at close range the large-scale irrigation project which had been built at Riverside with its concrete-lined canals—one of the first projects in the West to use this type of canal construction. Below the canal, he noted, were orange groves worth one thousand dollars an acre; above the canal, the same land sold for $25 an acre. Then, for some years, Marshall was recalled to Washington, where he served in the U. S. Army Engineers with the rank of colonel.

Back in California in 1919, Colonel Marshall printed a series of pamphlets on the "Marshall Plan" for the development of the Central Valley and began to speak from one end of the state to the other for the plan. From an engineering point of view, the Marshall Plan was a rather naive conception: the good Colonel had two canals, large enough to carry sailboats, going down each side of the valley. The drawings of these river-sized canals with sailboats seem to have had an enormous fascination for the Californians. Experienced engineers scoffed at the plan but the Colonel was undaunted and managed to persuade the legislature, in 1921, to appropriate $200,000 to investigate the plan. It is interesting to note that the Marshall Plan was a personal, not an official report; the dream of one man. But, however naive it may have been as to detail, it did capture the imagination of the people. Colonel Marshall died on June 22, 1949, but until his death he lived in San Francisco and took the keenest interest in the realization of his life-work, the Central Valley Project. In his later years he spoke in a whisper because of an artificial larynx necessitated, he thought, by the hundreds of speeches he made up and down the valley in support of the Marshall Plan.

In 1921 a bill was introduced in the state legislature to give legislative approval to the Marshall Plan. Known as the Water and Power Act, this bill passed the Senate, but was killed in the Assembly by the narrow margin of 4 votes. Two assemblymen who were active in the fight to defeat this bill had interesting subsequent careers. Elmer Bromley became the chief lobbyist for the Pacific Gas & Electric Company, and Frank F. Merriam later became governor and, in that capacity, vetoed legislation favorable to the Central Valley Project and loaded the Public Utilities Commission with appointees recommended by the private power interests. Once the Water and Power Act was defeated in the legislature, a group of wealthy California progressives, including William Kent, Dr. John R. Haynes, James D. Phelan, and the indomitable Rudolph Spreckels, took the issue directly to the people in three successive initiative campaigns in 1922, 1924, and 1926; but each time the proposal was defeated. In the first campaign, the Pacific Gas & Electric Company alone spent over $500,000 to defeat the bill by comparison with a total of $156,990.05 spent in support of the measure. Every dollar that the P. G. & E. spent to defeat the bill was, of course, charged up as an expense of operation, and was recaptured in the form of higher charges for electric power. The story of these campaigns, told by Carl D. Thompson, is amazing in its every detail.[2] Although the campaigns were unsuccessful, they had a powerful effect in arousing public interest in the Central Valley Project; in drawing a sharp line between the opposing forces; and in exposing the private interests that manipulated various "front" organizations. That these campaigns were conducted in the early twenties, when the general current of political thought was anything but progressive, is some measure of the strength of purpose and the tenacity of the wealthy progressives who underwrote them.

Then, in 1930, came the State Water Plan: the first comprehensive inventory of the total water resources of the state. This report dealt with every phase of the water problem: floods, saline intrusion, power, irrigation—all the inter-related aspects of water conservation and development. The report also projected a plan for the development of the Central Valley. Coming in the wake of the stirring campaigns of the twenties, this report prepared the way for the adoption, by the state legislature, of the Central Valley Project Act of 1933 which provided for the issuance of 170,000,000 in revenue bonds. That this measure should ever have been approved by the reactionary state senate is only to be explained by the fact, emphasized by Robert de Roos in his excellent study of the Central Valley Project,[3] that the Senators had "their eyes cocked on the possibility of a federal handout." For by 1933, the depression was well-advanced and the federal government was on the lookout for large-scale public works projects. The private utility interests immediately launched a referendum on the act which, after a campaign as stirring as any of those in the twenties, was upheld by the voters on December 19, 1933, by a vote of 459,712 to 426,109. It should be emphasized that, in this campaign, the people ratified and approved the very policies which, of recent years, various interest-groups have been seeking to induce the

[2] *Confessions of the Power Trust* (1932), Carl D. Thompson.
[3] *The Thirsty Land* (1948), Robert de Roos.

Bureau of Reclamation to abandon. The vote on the referendum was also most revealing for it brought out the sharp sectional rivalries in California. Los Angeles County, having its own independent sources of water, voted against the measure by a margin of 2 to 1. About 70 per cent of the Central Valley's production goes to San Francisco for marketing and distribution.[4]

After the Central Valley Project had been upheld by the voters, the state made no effort to issue or dispose of the $170,000,000 in revenue bonds which were authorized by the act. Here, then, was an anomalous situation in which a state had boldly launched a project which it was admittedly unable to finance. But the "boldness" paid off, for, in August, 1935, President Roosevelt authorized an initial appropriation of $12,000,000 to launch the project. Once this initial appropriation was made, the federal government was gradually "eased into" the project, step by step, with the whole question of control being more or less avoided by both the state and the federal government. During these negotiations, the state officials consistently evaded the question of whether this was to be a state project with federal aid or a federal project with state assistance. At this time, the state was not concerned with the question of ultimate control. It was concerned with federal aid. In repeated hearings in Washington, spokesmen for the state kept insisting that the matter of control was purely "secondary" and frankly admitted that California was unable to finance the project.

GIANT OF GIANTS

TVA and Grand Coulee are simple, a-b-c engineering projects by comparison with the Central Valley Project. As currently projected, the Central Valley Project will cost more than two billion dollars when it is completed. For what has finally emerged from 75 years of planning, controversy, and conflict, is a scheme which will provide an integrated system of stream control and irrigation works for every stream entering the valley. In short, the project is based upon a *total* utilization of every available source of water in the area. The project is far too complex, from an engineering

[4] Fresno *Bee*, May 20, 1949, p. 6-B.

point of view, to be described here; suffice it to say that, if the project is completed in accordance with the master plan worked out by the Bureau of Reclamation, it will make irrigation water available, as a supplement to present supplies, for 2,000,000 acres of valuable farm land; bring 3,000,000 acres into production; generate some 7,000,000,000 kilowatt-hours of power; improve navigation on the Sacramento River; supply water and power to many municipalities; protect some 300,000 acres of valuable delta lands against saline intrusion; store 30,000,000 acre-feet of water in 38 reservoirs; and add to the economic resources of the state to the tune of about $275,000,000 a year. Beyond all doubt, the Central Valley Project is the largest single development scheme ever undertaken in this country and, today, it is well on the road to early completion (the first Central Valley water was delivered in Fresno on April 29, 1949). One can say without exaggeration that the full and early realization of this project is absolutely vital to the future well-being of the people of California.

Considering the importance of the project and realizing that it is being federally financed, it would seem almost inexplicable that the project should still be involved in a most intense and bitter controversy; but such is the case. The diehards who have opposed the project for nearly seventy years have not abandoned their opposition, although the basis of this opposition has constantly shifted. Essentially the fight now turns on the question of *control*. The opposition concedes, in fact, it assumes, that the project will be completed; but it has never abandoned the fight for control. The question of control, moreover, is vital for the social values implicit in the project hinge upon administrative policies which in turn raise crucial questions: Who is to control the project? for what ends and purposes? in whose interest? At the present time, these questions involve three major issues: the 160-acre limitation; the question of power; and the issue of state vs. federal control. There are, of course, other areas of controversy; but these are the central issues.

Since its adoption in 1902, the Reclamation Act has contained a provision to the effect that, on all projects undertaken by the Bureau, landowners must agree to sell their "excess holdings," that is, holdings in excess of 160 acres in individual ownership (which has been construed to mean 320 acres under the community property laws of California) at prices which do not reflect the added value which would accrue after the completion of the project. As applied by the Bureau, this provision means: 1. That a landowner may retain *all* of his holdings if he elects *not* to use project water (he is under no absolute compulsion to sell); 2. That the landowner may retain all his holdings and still get project water for 320 acres (160 acres if the land is held in single ownership); 3. If the landowner, however, wants project water for his excess holdings, he must enter into a contract with the Bureau to sell his excess holdings, at a price set by an impartial board, within *ten* years from the date the contract for water is signed. The only compulsion in the act relates to the amount of *water* which any one landowner may obtain, not to the amount of land he may hold. The policy which this provision of the law reflects, namely, that individual owners of large tracts should not be permitted to monopolize the benefits which accrue from the expenditure of federal funds, is both clear, obvious, and equitable. From 1902 to the present time, the Bureau has taken the position that the Reclamation Act was passed, not so much to irrigate arid lands, as to make homes; its objectives, in other words, are social as well as economic.

In other areas of the West, this provision of the act has not met with active opposition primarily because, in these areas, most of the land to be benefited was in the public domain; but, since the historic American policy never operated in California, all of the lands affected by the Central Valley Project are in *private* ownership. Hence the opposition to the 160-acre provision in California is only to be understood in terms of the century-old effort to break-up large holdings. As might be expected, the *number* of landowners affected by the acreage limitation provision is not large; 90 per cent of the irrigated farms, by number, are held in units of 160-acres or less. In a study made in three key valley counties, 9,814 farms of a total of 12,941, were made up of units of 80 acres or less; 11,434 farms were made up of units of 160 acres or less; and 12,305 farms represented units of 320 acres or less. *But* this same study shows that 636 owners hold 53 per cent of all irrigable lands. Of these larger landowners, 18 had properties larger than 5,120 acres and, in the aggregate, owned more land than the 11,434 owners with units of 160 acres or less. Of 774,156 acres in the Central Valley area which have been surveyed by the Bureau, the owners of excess lands numbered 469 or 4.6 per cent of all ownerships. But abstract facts and figures fail to convey the reality of the acreage controversy or to indicate the character of the opposition. A brief glance at the history and background of one of the principal opponents to the acreage limitation provision will reveal more about this controversy than a volume of figures.

THE IRRELEVANT COW-COMPANY

Senator Sheridan Downey, who has been the leader in the fight to repeal the acreage limitation provision, published an interesting book in 1947 entitled *They Would Rule the Valley.* A section of this ghost-written volume is devoted to the thesis that the Kern County Land Company is, in relation to the acreage limitation provision, merely an "irrelevant cow-company." Let's see, first, whether this company is really just a "cow-company" and, second, whether its opposition to acreage limitation is really "irrelevant."

The history of this one company might well be a history of the Lower San Joaquin Valley for it has played, from the earliest date, a dominant role in the affairs of this section of the Central Valley. The company dates back to 1874 when a curious gentleman by the name of James Ben Ali Haggin, of Turkish origin, born in Elmendorf, Kentucky, visited Kern County and decided to invest a portion of his immense Comstock Lode wealth in a land-and-cattle empire. Further north, near Sacramento, Haggin owned the famous Rancho Del Paso, the world's greatest thoroughbred race horse farm. For his Kern County scheme, Haggin formed a partnership with Lloyd Tevis, a wealthy San Francisco financier. Henry Miller, of the famous Miller & Lux firm, was earlier on the scene than Haggin and Tevis, and had acquired 100,000 acres along the San Joaquin and Kern Rivers, a portion of which ran for

50 miles along both sides of the Kern. He had also built a huge canal, 100 feet in width, which carried water a distance of some 50 miles to other properties that he owned. Haggin & Tevis went farther upstream and began to buy land along the headwaters of the Kern River and to file claims to the use of this water. In this manner the issues were joined in the historic battle between Miller & Lux and Haggin & Tevis; or, in legal parlance, the battle between "riparian" rights as against the "doctrine of appropriation." The later doctrine was based, in part, on California mining custom and practice. In the mining camps, priority of occupation, which validated a mining claim, gave rise to the idea that prior use or appropriation established the right to use water to work the claim. Henry Miller finally won the battle, but only after two hearings before the Supreme Court and after one member of the court did a curious flip-flop, which is a most interesting story in itself. Later, however, Haggin & Tevis made their peace with Henry Miller by agreeing, in 1888, to convert Buena Vista Lake into a 25,000-acre water-storage reservoir which was, at that time, the largest storage basin in the country. In effect, therefore, Miller got the water he wanted and Haggin & Tevis got two-thirds of the land.

Following this compromise, Haggin & Tevis began to acquire more and more land—in California, New Mexico, Oregon, Arizona, and Mexico so that, by 1890 when the Kern County Land Company was formed, their holdings totaled 1,369,576 acres. Fifty per cent of the stock of this company, incidentally, is still owned by the heirs and descendants of Haggin and Tevis. Most of the 413,500 acres which the company owns in California is located in Kern County and about 250,000 acres are located on the floor of the valley within the Central Valley Project area. But this does not adequately indicate the basis of the company's intense interest in the acreage limitation for the company is also engaged in the water business. It now owns, through subsidiaries, approximately 80 per cent of the total flow of the Kern River, about half of which is used to irrigate its own lands and the other half of which it sells to farmers. To distribute this water, the company has built 800 miles of canals; owns 14 canal companies; and sells water to more than 800 consumers. Back in 1877, Haggin stated that he did not intend to monopolize land but that, as soon as irrigation systems had been extended, he would subdivide his holdings in small tracts. The promise has yet to be redeemed. In fact, the company added 4,000 acres to its California holdings in 1933. From 1907 to the present time, the company has steadfastly refused aid from the Bureau of Reclamation in the development of Kern River projects, and has blocked numerous water and power developments. The all-powerful position of the company in the affairs of the county is, in fact, frankly admitted by Senator Downey who has stated that "in the Kern County area the project simply cannot get to first base without the cooperation of the company."

Now, a brief glance at the operations of the company. For years the company has raised cattle on its out-of-state holdings for feeding and fattening in California. It normally carries about 70,000 head of cattle and its cattle operations are thoroughly integrated. In 1936 oil was discovered on the company's land so that today some 500 producing wells are located on various portions of its holdings. In listing 2,000,000 shares of its capital stock on the New York, San Francisco, and Los Angeles exchanges in 1948, the company reported a net income for the year of $9,465,264 and gross earnings of more than $17,000,000. The "relevance" of this company's opposition to the acreage limitation should, therefore, be fairly obvious. If the acreage limitation is retained, the company will be forced, if it elects to use project water, to sell thousands of acres of excess holdings without being able to reap the king's ransom that would come from the added value which project water would confer on these lands. Part of the project involves the construction of dams on the Kern River, and if these dams are built under the Reclamation Act, the company's lands, irrigated and unirrigated, will be subject to the acreage-limitation provision.

The importance of the acreage limitation is that it represents a device by which, after a hundred years, the pattern of large land holdings in the valley can be changed. Ultimately the pressure of population will force a subdivision of these holdings but the question remains: must the people of California pay an exorbitant premium for these lands, representing the added value conferred by the project, or shall these lands be placed on the market, within the next ten years, at a price which does not reflect this unearned increment? In terms of public policy, there can be little doubt as to what the answer to this question should be. It can hardly be contended that the people of the 48 states appropriated millions of dollars to build a project in California for the special benefit of 4 per cent of the farmers of the state.

The acreage-limitation provision, as it applies to California, is complicated by the problem of underground waters. Underlying the lower San Joaquin Valley are some 20,000,000 acre-feet of usable storage space. The waters in this underground storage are made up of "geological" waters; the annual run-off; and waters from normal irrigation diversion less evaporation. Nowadays some 6,000,000 acre-feet of water are pumped annually from this underground supply, principally during the months when the surface-waters are inadequate. Once Central Valley water is made available, the underground sources will be "recharged" and the water level is likely to rise. There is, of course, no way by which project water can be distinguished from non-project water in the underground basins. Under California law, a landowner can use the water that percolates beneath his land, but prior use, in this case, is not a factor. Hence a large landowner who decided not to use project water could continue to pump from the underground supply which would be augmented by the increased seepage and drainage, which would take place once project water was available. It is argued, therefore, that a recalcitrant landowner might profit in this manner without sharing in the cost of the project. For, as the amount of the underground supply increased, the water level would rise and pumping costs would decline.

Although this is an important issue, the state can always regulate the right to pump underground waters; some states already have done so and there is no doubt that California needs legislation of this character. Even in the absence of legislation, the California courts might be induced to place some limitations upon the right to pump underground waters. In the second place, the large landowners are eager to receive project water no matter how noisily they proclaim their lack of interest. The water shortage is such, moreover, that it will be a long time, indeed, before the water from the little farms in the area will again fill the underground pool. Besides, the underground supplies are not evenly distributed throughout the area; there are many subterranean walls and barriers. In the last analysis, the demand for water is almost certain to reach a point where, even with project water available, the supply will be slightly deficient.

The real "loophole" by which the large landowners may escape from the acreage limitation is to be found in the fact that they can make a token compliance with the provision while continuing to operate their holdings as a unit. Here the possibilities are endless. To be sure, the Reclamation Act stipulates that the landowner must be a bona fide resident on the land or "occupant thereof residing in the neighborhood." But the large landowners have large families and could doubtless find many sisters, cousins, and aunts who might become residents or occupants; and there is nothing to prevent large leasing operations. Despite the possibilities of a legalistic sabotage, the acreage limitation can be used, in connection with the steadily mounting pressure of population on land resources, to break up the large holdings. The question of farm labor, also, has a direct bearing on the outcome. Part of the alleged economies of large-scale operations are to be accounted for in terms of the disadvantaged position of farm labor. If the recommendations of the La Follette Committee were to be adopted, the large operators would no longer be able to tap a large, unorganized, supply of cheap labor and some of the competitive disadvantages of the small farm would disappear.

THE POWER QUESTION

Closely related to acreage limitation is the question of public versus private distribution of the power to be generated by the project. This has ceased to be a question of whether the power should or should not be generated; the "power famine" of 1948 settled this issue. But the question remains: shall the power be distributed by the Pacific Gas & Electric Company or by public agencies? Actually there is just one company involved: the Pacific Gas & Electric, a giant utility empire formed by the consolidation of some 449 water-and-power companies. The North American Company, one of the nation's largest holding companies, owns 31.98 per cent of the common stock of P. G. & E. and the second largest stockholder is Standard Gas & Electric Company of New York. The area in which the P. G. & E. today enjoys an unchallenged monopoly in the distribution and sale of power embraces virtually all of northern and central California. The political power of the company, needless to say, is proportionate to the extent of its monopoly. It is now and always has been knee-deep in California politics. Perhaps the extent of its power might be indicated by the statement that it controls sources of energy equal to the labor of 50,000,000 men.

The battle over control of Central Valley power is so highly dynamic that it would serve no purpose to define the specific issues here; by the time this manuscript is published the issues will probably have changed. Suffice it to say that the P. G. & E. has been notably successful in sabotaging the power potential of the project: first, by contending that there was a "surplus" of power in California; and, later, by depriving the Bureau of Reclamation of funds to build transmission lines and stand-by plants and, also, by sewing up the available markets for power. Since about one-third of the power generated by the project will be needed to operate the project's pumping plants, the company has been able to develop a specious argument for the benefit of farmers. The company has offered to buy power from the Bureau at what would appear to be a fair price and to provide power required for pumping and other project needs on an exchange basis. This offer, so the argument runs, would bring in more revenue and thereby reduce the charge for water. But there are many obvious "catches" in the offer. It includes, for example, a charge by the company of 5 mills per kilowatt hour for the service rendered in carrying project power from the government's dams to the government's pumps and this charge would run indefinitely. Furthermore, one-third of the power will be sold to irrigation districts and farmers for pumping and the remaining third will be sold to other consumers. Thus the farmer is also a "consumer" of power as well as water, and cannot hope to benefit from the company's offer.

What the power question involves, therefore, is the issue of control with all that is implied in this statement. Regardless of how effective a delaying action the company can conduct, the federal government will never acquiesce in a situation in which control of a two-billion-dollar project is turned over to a privately owned concern. It is the control factor, not the ownership factor, per se, that is important. To realize the full social and economic benefits of Central Valley power, the government must control the sale and distribution of this power from the generators to the ultimate consumers. Although the earnings of the P. G. & E. are regulated by a state agency, this regulation is in terms of the capital stock structure of the company. With government operations, on the other hand, power rates can be determined, not on the basis of what must be paid preferred stock owners but, once the initial costs are repaid, on the basis of operating and maintenance costs. This difference, from a long-range point of view, is enormous, and constitutes the ultimate argument for public ownership.

Sooner or later the state or the federal government must take over the P. G. & E. lock, stock, and barrel. There is good reason to believe that the astute executives who manage the company are not only fully aware of this possibility, but that their present strategy is based on the assumption that the company will some day be taken over by a public agency. The aim, therefore, is to delay public acquisition as long as possible and, at the same time, build up a book value which the government will some day have to pay for the company's properties. By tying-up contracts for the sale of power, by fighting the Bureau of Reclamation every step of the way, and by continuing to subvert public opinion in California, the company can count on a fairly extended term of existence.

The Pacific Gas and Electric Company has important allies in California. There are those who believe that the acreage limitation fight might have been settled long before now had it not been for the power question. The P. G. & E. makes skilful use of the acreage limitation fight as part of its power strategy and the opponents of acreage limitation are, without exception, the allies of the company on the power issue. In the aggregate, these two sets of interests make a formidable opposition. The power of this opposition, moreover, has been augmented by the fact that a long and bitter feud has prevailed between the Army Engineers and the Bureau of Reclamation. The special interest groups have long used the Army Engineers to fight the Bureau. Thus the government has not been able, on all occasions, to present a solid front to these interests. The Army Engineers affect an air of "neutrality" but never seem averse to accepting support from or collaborating with the most reactionary interests in California. The engineers are interested in flood control and navigation—their special vested interest—and have never been greatly concerned about multi-purpose projects. If power is to be generated on a project, well and good; but they will not fight for power. Their specialty is the construction of low-level single-purpose dams for flood control—the more dams the better. The existence of this internal feud within the federal service has greatly hampered those elements in California who have consistently fought for the idea of an integrated project, carefully articulated, comprehensive in scope.

STATE VS. FEDERAL CONTROL

Initiated by the State of California, taken over piece-meal by the federal government, the Central Valley Project is an administrative anomaly. Once it is completed, the question of ultimate control will become crucial. From 1933 to the present time, a powerful coterie of interest-groups in California has consistently urged that the federal government should turn over the entire project to the state. This is the position of the State Chamber of Commerce, the Pacific Gas & Electric Company, of Harrison Robinson of the Canners Institute, and of the Farm Bureau Federation. These interests have been able to induce the state legislature to go on record in favor of state control, and many boards of supervisors, including the Los Angeles County Board, have taken a similar position. Should the project be turned over to the state, it would mean that the legislature, by a mere majority vote, could so amend the Central Valley Project Act as to subvert its stated purposes. The Water Project Authority, which would assume control of the project, is made up entirely of ex-officio state officers, three of whom are chosen by election. Hence state control would constantly involve the project in state politics with a fair chance that special interest groups might gain control of a two-billion-dollar public enterprise by electing a majority of the members of the authority. Corporate interests would then control, indirectly, privileges not gained by or dependent upon any monopoly of individual initiative, but rather acquired by, and inherent in their superior capacity for manipulating state and local politics.[5]

Emphasizing the purely intrastate features of the Central Valley Project, the Los Angeles *Times* continues to campaign for state control.[6] But the plain fact is that, since 1933, the state as such has not functioned effectively in its particiption in the development of water resources and that the initiative has definitely passed to the federal government. The state legislature admitted as much in 1943 when, in a joint legislative report, it stated that "a new era of federal concern in state affairs has developed. . . . Whereas previously she has dealt with these problems on her own behalf, California now must provide the means by which the efforts of federal agencies to assist in the solution of these problems are coordinated with the best interests of the State as a whole." The state has no agency that is even equipped in theory to take over a project of this magnitude. The members of the State Water Project Authority consist of state officers elected or appointed primarily for other functions; nor has the authority functioned with much effectiveness for the last decade.

One thing is clear: the project must be administered by a single agency if for no other reason than because the exchange of waters on a basin-wide basis demands a central administration. To divide authority as between, say, the Bureau of Reclamation and the Army Engineers, would mean that there would be no real authority vested in either agency. "We have within one State," to

[5] *See:* Articles by Dr. Arthur D. Angel in the San Francisco *News,* Jan. 22 to Feb. 2, 1945.

[6] Los Angeles *Times,* June 5, 1945.

quote Richard Boke, the able Regional Director of the Bureau, "problems of a complexity and variation that would stagger a nation; areas of high industrial development and areas not far advanced from pioneer days; rich farm lands and poor; wide range lands and rolling dry farm lands. A program designed to benefit all these areas and their people can not be simple and it can not be piecemeal. It must be integrated. Not one canal should be built without consideration of its effect on other installations and needs in the state."

It has been suggested that there should be an authority established, after the pattern of TVA, that would cover the drainage basin of all streams flowing into the Pacific south of the Oregon-California boundary. But there are some difficulties involved in the application of the TVA pattern in California. First of all, more than one watershed is involved in the Central Valley Project. In this case, moreover, the authority proposal comes late in the stage of proceedings, for the project is now well on the way to completion. In the Tennessee Valley, the TVA filled a vacuum; but California had a highly developed water program long before the Central Valley Project was authorized. There are over a hundred irrigation districts organized under state laws as public agencies with approximately 6,000,000 acres of land, one-half of which is already under irrigation. Despite the difficulties, however, a decentralized authority patterned after TVA would seem to be the best long-range answer to the problem. Obvious perils are involved in permitting a project of this magnitude to remain, for an indefinite period of time, under the control of an agency like the Bureau of Reclamation which, like all similar agencies, can have its functions enlarged or impaired by a simple act of Congress. An authority would seem to have a better chance of resisting sectional pressures and conflicting interest-groups than a federal agency that has to run to Congress for additional legislation in order to meet new problems and issues as these arise.

[18]

O PANTHER OF THE SPLENDID HIDE!

◇◇

> How *art thou conquered, tamed in all the pride*
> *Of savage beauty still!*
> *How brought, O panther of the splendid hide,*
> *To know thy master's will!*
> —BAYARD TAYLOR

EARLIER CHAPTERS of this book have dealt with various California problems; this final chapter will be concerned with *the problem of California.* Is there, then, a California Problem and, if so, what makes California a problem? The emergence of a new center of social, economic, and political power, in any area, will invariably upset the antecedent balance of power and disturb the relationships based on the prior equilibrium. The truism of international politics also applies, of course, to the balance of power between regions and states. In the last hundred years, the gradual emergence of a new center of power in California has had a constantly disturbing effect on inter-state and inter-regional relationships and has brought into being a peculiar relationship, often verging on outright antagonism, between California and the federal government. California happens to occupy a geographically marginal or peripheral relation to the other western states and the nation, a circumstance that aggravates the disturbing effects. In this case, also, the new center of power has emerged so recently, in historical terms, that not enough time has yet elapsed to develop a new pattern of relationships. Developing *outside,* and to a degree independently of, the other western states, and being a more or less self-contained regional unit, California has become a major problem not only to the western states but to the nation.

CALIFORNIA: *The Great Exception*

THE PROBLEM OF CALIFORNIA

One measure of California's power is to be found in the social and economic vacuum which, until recent years, existed in the Inter-Mountain West. Historically, there has been no counterforce in this vast area to balance, steady, or check the power of California. In 1860, when the census first reported populations in the new Pacific and Mountain Divisions, the former had a population of 105,891, the latter of 72,927. But, in the decade from 1850 to 1860, the population of the Pacific Division increased 322 per cent, that of the Mountain Division by only 144 per cent. In the next four decades, however, the Mountain Division showed a more rapid rate of increase than the Pacific Division, largely as a result of the eastward expansion of the mining frontier from California. In 1900 the balance of population stood at 2,416,692 for the Pacific Division; 1,674,657 for the Mountain Division. Then the second great "leap-frog" movement of population to the Pacific Coast began so that today the Pacific Division has 14 million population, the Mountain Division, 4.4 million. To set these figures in proper perspective, however, it should be kept in mind that most of the increase in the Pacific Division has been in one state, California, which today has more than half the total population of the 11 western states.

As a chauvinistic westerner, I have participated in any number of projects, during the last two decades, which had as their primary purpose the stimulation of a heightened western regional consciousness. Extensive use was made, in most of these political ventures, of the familiar theme that the West is a colonial dependency of the East. Over a period of time, however, it gradually dawned on me that this theme seemed to strike a more responsive chord in certain parts of the West than in other parts. It appeared to make sense to the people of Montana, Colorado, Wyoming, Utah, and New Mexico; but, for some reason, the Californians always declined the bait. In fact, more than one ambitiously projected "western states conference" has been still-born because of California's lack of interest and enthusiasm. Obviously

the Californians do not think of themselves as "colonials" no matter how loudly the theme is proclaimed.

I finally came to the conclusion, therefore, that the theme of the
West as a colonial empire had to be taken with a grain of salt. It
was unquestionably a sound theme in the Inter-Mountain West,
which still carries the visible scars of Eastern exploitation; but one
can look in vain for these scars in California. The more I thought
about the problem, the more I came to see that California occupies
a relation to other parts of the West which is comparable to the
relation that the East occupies to the West. The Inter-Mountain
West is a colony in a two-fold sense: a colony of the East *and* a
colony of California; whereas California is a colony of the East in
only the most limited sense. I then began to see how successful
California has been in using the theme of "colonies" and "colonial
status" as a smoke-screen to conceal certain important aspects of its
relations with the other western states.

The difficulty with our social vision has been that an all-inclusive
phrase like "The West" conceals the discrepancy of power which
actually exists between the western states and California. By the
end of 1945, California and Washington had 5.91 per cent of the
manufacturing employees of the nation as compared with 1.53 per
cent for the remaining 9 western states. Actually the bulk of these
manufacturing employees—4.76 per cent—were in California. In
1943 the 11 western states had 14.77 per cent of national income
payments but, of this total, 8.67 per cent was accounted for by
California alone: more than half the total. As industrial power is
measured nowadays, California is still at a disadvantage by comparison with older and industrially more mature regions but *in
intra-regional competition* it is clearly dominant. For many years,
California has been in a much stronger financial position than the
other western states and has invested large surpluses of capital in
various enterprises in these states. *Within the West,* California has
long enjoyed a favorable position even with respect to freight rates
since the availability of ocean transport has compelled the railroads
to give preferential rates to the terminal points on the coast. In
relation to eastern manufacturers, therefore, California manufacturers have a favorable rate throughout the entire area extending

from the coast as far east as the Rockies. Essentially, California is to the West what New York, for many years, was to the industrial East: a great center of power with lines of influence radiating outward in all directions. But there is this important difference: California, within the West, has no rivals.

In addition to its unique problems, California has all the problems of all the other western states. But since it is socially and economically more mature than these states, with greater power and "the momentum of an early start," it undertook the solution of many of these problems on its own initiative, apart from the other western states, and without the aid of the federal government. In this respect, its marginal position and its exceptional advantages made possible a development-in-isolation which gave the state a greatly enhanced sense of its own power and independence. When California has gone out for anything, a project, federal funds, or whatever, it has usually acted not as one of 11 western states but as a nation demanding what it had the power to take and the record will show that it has been highly successful in these unilateral raids and political maneuvers. As long as the development of the other western states was retarded, California could act in this imperial manner. But today, with the West coming to a new maturity, California is rapidly discovering that its interests are closely related to the interests of the entire region. However, California has yet to develop a full awareness that it has now entered upon a new phase of its development. It is the failure to realize this fact, indeed, which makes up the California problem. The crucial and immediate question, therefore, is whether this "panther of the splendid hide" can come to know its master's will before disaster overtakes it.

THE END OF EXCEPTIONALISM

On March 30, 1949, the number of unemployed in California stood at 528,000, an unemployment rate of 14 per cent: twice that of the nation as a whole. One-fifth of the nation's unemployed, for the first quarter of 1949, were to be found in the three west coast states. Calling attention to this fact in a speech in Los Angeles on May 11th, Governor Earl Warren imparted the further bad news that California must prepare to receive "another 10,-000,000" migrants in the next quarter century. Faced with the highest unemployment rate in the nation and with the expectation of continued large-scale migration, California, the Governor went on to say, must undertake a rapid development of all available resources and, in particular, must increase the sources of available energy. "California," he said, "has no great coal fields. . . . At the rate of present use, our known oil and gas reserves will last only about 20 years." Where, then, is this additional energy, so vital to the further expansion of California, to be found?

With the exception of federal projects on the Colorado, California's power program is largely intra-state for its river basins are practically all confined within its own borders. Although there is a large undeveloped energy potential within the state, most of the suitable sites for upstream run-of-the-river hydro systems are already fully developed and further expansion is unlikely. The undeveloped potential, therefore, involves river basin developments. But these developments are inseparably connected with water conservation for all purposes, including irrigation, flood control, navigation, salinity control, recreation, municipal uses, and so forth. Since none of the existing private utility systems has the resources to undertake developments of this character, California has been driven, and will continue to be driven, to seek federal support. Federal support, in turn, hinges on the goodwill of the other western states which have few congressmen but many senators.

Even assuming continued federal support, however, California cannot rely upon its own resources; in fact, it is already directly dependent upon one inter-state resource, namely, the Colorado River. Speaking at the State Water Conference in 1945, Leland Olds, of the Federal Power Commission, summed up California's predicament when he said that the state's energy resources will prove inadequate for the full development of a balanced economy if California's program is conceived entirely in intrastate terms. "The fact is," he said, "that unless your water resource program here in California is recognized as an integral part of the larger program which embraces the entire Pacific and intermountain region, your

[*345*]

future development may ultimately be restricted by lack of energy resources . . . Your planning cannot safely assume that California will remain self-contained so far as its energy requirements are concerned." Fortunately the West, as a region, has a developed capacity of about 9,000,000 kilowatts of hydro-and-steam electric power; but, with full development of all resources, it can generate an additional 50,000,000 kilowatts of power. It is this power and this water in which California must share if it is to support its present and anticipated population. Failing to recognize this long-range dependency, Calfornia continues to act as though it were the only state west of the Rockies. In doing so, it is holding up the development of the West and also imperiling its own and the nation's economy.

By the time the Central Valley Project is completed—assuming that it is completed as planned—the federal government will have an investment in this one California project of two billion dollars: two and one-half times the federal investment in TVA and almost twice as much as the Bureau of Reclamation has invested in all the other western states in the 43 years of its existence. With an investment of this magnitude, the federal government is bound to "intervene" in California affairs and to insist upon the observance, within the state, of nationally approved policies. At the same time, the other western states, now developing similar projects, feel that California has too long received the lion's share of federal subsidy and "view with alarm" California's attempt to monopolize water resources. "If the interests of the nation as a whole are overlooked by the people of California," warned Mr. Harry W. Bashore at the State Water Conference, "I believe I can say without qualification that you will make little progress with Congress in obtaining the funds you so urgently need." But the problem involves more than placating the other Western states: the fact is that California's resources are basically inadequate to support its future population. "Remember," said Mr. Bashore, "that the day is not far distant when the population of your state will be greater than your resources can sustain, if you do not take wise action now. As population increases, wise husbanding of resources is necessary if you wish to escape the ultimate fate of almost every civilization the world has ever known. . . . Likewise the population of California will increase beyond your ability to support them and your own prosperity will decay, unless you give heed in time to developing your resources to their maximum long-term capacity. Piecemeal development, project by project, will not accomplish this. Only an overall plan which is sound will serve this end." In this sense, an overall plan means, of course, a plan for the entire West, not for California alone.

JACK AND JILL

Jack and Jill had only to go up the hill to fetch a pail of water, but California may have to go to the Columbia River for water and the highway measures 1,023 miles from Los Angeles to Portland. In 1948, when California was suffering from a drouth, Oregon was experiencing one of the worst floods in its history. "Two months after California was seared by a crippling drouth," writes Richard Neuberger, "Oregon and Washington and the Idaho 'panhandle' wallowed in a disastrous flood." As Californians scanned the skies in search of rain, Oregonians prayed that the flood waters would abate.

Although the possibility of diverting Columbia River water to California may sound fantastic, the fact is that the Bureau of Reclamation is already studying the feasibility of such a project. This study has crystallized in what is known as the "United Western Plan." For years engineers have dreamed of some such plan whereby the 160,000,000 acre-feet of Columbia River water which waste into the Pacific every year might be utilized. The engineers believe that water could be diverted from the Snake River, one of the tributaries of the Columbia, into the Colorado near its source, or that water could be diverted from the Columbia at a point below Bonneville Dam so as not to disturb any present use of the river. Actually the Columbia discharges more waste water into the Pacific every year than is to be found in all the other western rivers combined, including the Sacramento and the Colorado. After making allowance for all present consumptive uses, this water would irrigate 40,000,000 acres of land: more than twice the amount now in irrigation throughout the United States. According to Mr. William E. Warne, Assistant Secretary of the Department of Interior, the United Western Plan is entirely feasible, and the realization of this plan or some version of it is to be anticipated in the next 20 or 30 years. There is a vast acreage of hard, caked, sage-brush land in eastern Nevada which, with water, could grow peaches, asparagus, alfalfa, and many other crops. The general feasibility of the plan is indicated by the fact that much of California and the intermountain region has only 10 inches of rainfall annually, whereas the Columbia lowlands are frequently saturated with as much as 100 inches of annual rainfall. After all, the Columbia diversion plan would represent merely an extension of the same principle which is now being used in the Central Valley Project where water is being diverted a distance of 500 miles from one watershed to another.

Although suffering from periodic floods, the Pacific Northwest, needless to say, does not look with favor on the plan to divert Columbia River water to California. "If this fantastic project goes through," remarks Senator Guy Cordon of Oregon, "Oregon would like to have a corresponding conduit going north, pulling into it the income of California oil plus the income of Hollywood. If a few stars fall into the sluiceways, we'll gladly accept them." In a similar vein, the Portland *Oregonian* raises the interesting question: "Why should not the people come to the water, instead of the water being transported—at an initial cost of possibly $1,000,000,000—to the people? There are no barriers of which we are aware to the migration of drouth refugees to the irrigable lands of Oregon and Washington, which are within easy reach of the great Columbia."

Although the need for Columbia River water in California may be debatable, there can be no question about the need for Columbia River power. Many years ago, James D. Ross, first Bonneville administrator, saw the ultimate necessity for a western power pool. Recently California congressmen have proposed that an interconnecting "grid" system should be arranged by which Bonneville power could be used in California. This, too, is an entirely feasible project, in theory, since California has need for power during the off-peak power demands in the Pacific Northwest. It has been suggested, for example, that the interexchange of power could be made by shuttling power to California between midnight and 6 A.M. when the demand is low in the Pacific Northwest, producing an estimated additional revenue for Bonneville of $3,000,000 a year. Conversely, if California had off-peak Bonneville power it could save $6,000,000 a year which is now spent on fuel oil for standby steam plants. An inter-connecting grid system would have great value in case of a national emergency, such as war, or natural disasters, such as earthquakes. It was estimated in 1948 that Bonneville could make available 1,570,000,000 kilowatt-hours of off-peak power to California. Bonneville, it should be remembered, is only 200 miles from Shasta Dam, so that the inter-connection could be made today for an estimated cost of $6,000,000, a figure equivalent to one year's savings on fuel in California. The grid systems of Oregon and California, moreover, are growing closer together through the process of population expansion, so that by 1952 the problem of inter-connection will be even simpler than it is today. If Bonneville and Shasta Dam were inter-connected, it would then be possible to integrate this system with that of Boulder Dam. An all-western grid system, in any case, would be the counterpart of the United Western Water Plan.

The "catch" in this dream-like proposal, of course, is "politics." In the first place, and strange as it may seem, Southern California is strenuously opposed to the United Western Plan for it fears that this plan was conjured up by the Bureau of Reclamation to divert California's pressure on Arizona in the feud over Colorado River water.[1] Congressman Richard J. Welch of Northern California suggested that perhaps the answer to Southern California's water problem was to be found in the Columbia Basin. Made at a time when Southern California was in the thick of its battle with Arizona, Congressman Welch's suggestion was furiously denounced in Los Angeles as "the kiss of Judas;" this "San Franciscan," it was said, "is trying not to succor but to sucker us." To suggest that Colorado River water was not the only water which might be made available in Southern California was, of course, an act of

[1] *See:* Los Angeles *Daily News*, Dec. 23, 1948.

treason, a betrayal. As a matter of fact, the Republican Party in California has been considering the possibility of ex-communicating Congressman Welch for this and other heresies.

Faced with the serious likelihood that it may have to go, hat in hand, to the Pacific Northwest for water or power or both, California has done nothing to cultivate better inter-state relations with Oregon and Washington; on the contrary, it has tried, on more than one occasion, to undercut their development programs. Concerned over cheap power rates in the Pacific Northwest, which attract industries like a magnet, California congressmen have been conducting a "cloakroom campaign" to increase the Bonneville wholesale rate for power so as to offset the competitive disadvantage of higher rates in California. The Portland *Oregonian* recently pointed out that "Power-starved, water-short California needs to get out of the habit of thinking of that state as a self-contained empire. It needs the Northwest as much as the Northwest needs California." The same editorial went on to observe that California "has been guilty of legislative and other efforts to cripple development of the Columbia Basin's hydroelectric and irrigation resources,"—a charge that happens to be well-substantiated.

Although one can easily understand the Pacific Northwest's reluctance to share water and power with California, the Northwest is highly dependent, nevertheless, on California in other respects. California markets are of prime importance to the Northwest, for the economies of the two regions are essentially supplementary rather than competitive. Oregon sawmills shut down when there is a lack of power in California and Oregon's range lands are invariably invaded when a drouth occurs in the San Joaquin Valley.[2] Despite this mutual dependency, however, it would be difficult to imagine a thornier political problem than that presented by the United Western Plan. In view of the 27-year-old dispute between California and Arizona, who would care to estimate the length and bitterness of the feud that might develop between Oregon and California over the water and power of the Columbia River?

Nor should it be assumed that the nation is or can ever be neu-

[2] *See:* "Influx of California Cattle on Ranges to Create Big Problem for Oregon," Portland *Oregonian*, March 14, 1948.

tral in issues of this kind. Bureau of Reclamation officials have attempted to by-pass the California-Arizona controversy by saying that it is merely "a family feud" to which the federal government is not a party; but this is far from being true. Western lands irrigated by reclamation projects produce $580,000,000 annually in agricultural products, all of which is a contribution to the nation's wealth; and eastern manufacturers furnish the lion's share of the materials used in western development projects, such as, generators, turbines, transmission lines, and so forth. Furthermore, the Columbia River has a potential power output equal to one-fifth of the U. S. coal production or one-fourth of the U. S. oil production. The more power that can be developed in the West, therefore, the less the drain on irreplaceable natural resources.

Once the drouth of 1948 was well-advanced, California began to realize the consequences of having neglected its relations with the other western states. Montana, for example, turned a cold shoulder on California's request for the use of pasture lands for drouth-stricken cattle. Colorado, also, was lukewarm to a similar proposal. In the midst of this emergency, California proceeded to "steal" Charles L. Patterson, the chief engineer of the Colorado Water Conservation Board, a man who had gained an intimate knowledge of Colorado's water secrets. Governor W. Lee Knous immediately protested to Governor Earl Warren that the hiring of this expert was highly "unethical." The Denver *Post,* in an editorial of February 29, 1948, commented that "Californians will use any means, fair or foul, in their plans to grab water from other states." "California water strategists," the editorial went on to say, "proceed on the assumption that so great a state is superior to the codes of upright dealing to which decent men everywhere subscribe." That California was able to "steal" this expert is, of course, merely further evidence of its greater financial power.

As a matter of fact, about the only western state with which California has cordial relations is Nevada, and Nevada is simply a satellite of California. The basic explanation of this relationship, moreover, is to be found in the fact that Reno and other Nevada communities on the lower levels of the Truckee River get most of their water from Lake Tahoe, "the lake of the skies," in California; also the Sierra Pacific Power Company, which distributes all the power in the Reno area, buys most of its power from the Pacific Gas & Electric Company. Like most satellite states, therefore, Nevada is really California's unwilling captive.

Nor is it merely water and power, vital as these are, that are involved in the western scramble for resources. Colorado is deeply concerned over the plans of the Pacific Gas & Electric Company to obtain prior and exclusive control of the natural gas from the vast, newly discovered San Juan oil dome in southwestern Colorado. At the present time, the P. G. & E. proposes to transport this gas by a 990-mile pipe-line to San Francisco; but Colorado has entered a vehement protest and the issue is still in doubt. Colorado is also concerned by the way in which California companies, notably Union Oil Company, have moved in on the Department of Interior's shale-oil development in northwestern Colorado which has a potential 200 million barrels of oil with probably another 100 million barrels in nearby shale deposits. At first it was feared that the development of these shale-oil deposits would be impossible because of the effect it would have on the water supply. But the oil companies have now developed what is called "an ultrasonic congolomerator" by which shale deposits can be processed without water and at costs which can meet the competition of natural oil. Recently, the states of New Mexico, Montana, Wyoming, and Colorado voted to form an organization whose goal would be the establishment of a western wool processing and fabricating system. At the present time, wool is processed almost entirely in New England, although 75 per cent of the wool is produced west of the Missouri River. In the absence of even testing plants, western wool growers must ship uncleaned wool 2,000 miles to market, and 60 per cent of the freight bill is accounted for by grease and dirt in the wool. It should be noted, however, that California did not receive a bid to this conference, although Nevada and Utah were later brought into the scheme. Obviously the other states fear California's superior power. These are but a few of dozens of similar illustrations that might be cited to make the point that where power is so evenly distributed, as it is between California and the other western states, cooperation becomes most difficult.

O Panther of the Splendid Hide!

CALIFORNIA THE TIGER

> There was a young lady of Niger
> Who smiled as she rode on a tiger;
>   They came back from the ride
>   With the lady inside
> And the smile on the face of the tiger.

The problems that the other western states and the federal government face in dealing with California cannot be better illustrated than by reference to the future plans which the Bureau of Reclamation has developed for the control and use of the Colorado River. In its great report of June 6, 1946, the Bureau outlined a plan calling for the construction of 134 projects or units of projects on the Colorado: 100 in the Upper Basin, 34 in the Lower Basin, at an estimated cost of $2,185,552,000. Fully developed, these projects would benefit 2,656,230 acres of land, bringing water to 1,533,960 acres not now under irrigation, and would add 3,500,00 kilowatts of power to the supplies of the southwest. It is estimated that these projects—with such fabulous names as Deadman Bench, Minnie Maud, Troublesome, Desolation Canyon, Roan Creek, Cochetopa Creek, and Yellow Jacket—would bring about an increase of gross crop income of $65,000,000 a year at pre-war prices and expand the tax base from 50 to 100 million dollars. The completion of these projects would mean new croplands; increased yields; better pasture; fatter cattle; more power; better flood control; new recreational facilities; more industries; larger bank deposits; more tax revenue. The realization of this plan, in the perhaps too-fervid phrase of William E. Warne, would bring into being a civilization along the Colorado as great as those that grew up along the Nile, Tigris, and Euphrates rivers.

But who is to say which of these 134 projects will be developed, which shall not? What standards shall so determine? What are the economic implications? Where stands the law and equity? Who is to develop these projects, in what order, and for what purposes? Above all, who is to control the projects once completed? To realize the importance of these issues, as seen by the other

[353]

western states, two facts must be kept in mind: first, that California is the only western state which has already constructed the basic works by which it can utilize Colorado River water; and, second, California gets "the last crack" at the river before it crosses the border. There are, therefore, precisely 134 potential disputes involved in this program. It should be kept in mind, in this connection, that California, or those who speak in its name, have denounced the entire plan as lacking in economic feasibility and as being "socialistic." [3] There can be no doubt, moreover, but that California will use the full weight of its power to retard and, if possible to sabotage this plan and that it will oppose, item by item, each separate project.

The Bureau of Reclamation has frankly stated that it cannot go forward with the development of the Colorado until some agreement has been reached among the states involved. No agreement, of course, has been reached between the states of the Lower Basin. However, the Upper Basin states, where only about one-third of the water is being currently used, have reached an agreement on the distribution of their share. On July 31, 1948, these Upper Basin states, meeting at Vernal, Utah, finally settled their differences and President Truman signed, on April 7, 1949, the compact which these states and the United States Senate had approved. One reason why this agreement was reached is simply that California, not being in the Upper Basin, did not have to be consulted. Much as she would like to have been a party to these negotiations, California did not participate. More than anything else, ironically enough, it was the fear of California's intention, held by the Upper Basin states, that brought about the agreement.

Through a "phony" organization called the Colorado River Water Users Association, which it strictly controls, California did everything in its power to prevent the Upper Basin states from reaching an agreement. Governor W. Lee Knous of Colorado denounced this association, which had called a rival conference at Salt Lake City, as a "stooge" organization set up by California to further its own selfish interests, and accused California of "foment-

[3] Los Angeles *Times*, June 25, 1947.

ing opposition to the ratification of the Colorado River Upper Basin Compact." [4] At the Salt Lake City conference, California spokesmen argued that the Colorado-Big Thompson project would "unbalance the natural division of the entire Colorado" and sought to rally the western slope counties of Colorado against the compact despite the fact that California is the greatest non-watershed user of Colorado River water. With the exception of California and Nevada, the other western states have long had their own organization: The Colorado River Basin States Committee in which, incidentally, Nevada and California refused to continue membership. Similarly, California looks with little favor on Mexico's attempts to develop such projects as the Morales Dam by which it could begin to make full use of the 1,500,000 acre-feet of Colorado River water which it has been guaranteed under the Mexican-U. S. water treaty of 1945.[5] In this respect, therefore, the California Problem is of direct concern to the federal government for it involves an important aspect of our foreign policy.

It is not surprising, therefore, that the Upper Basin states are Arizona's allies in the feud with California. Rightly or wrongly, these states believe that California's real aim is to upset the original Colorado River Compact and that, if necessary, California might repudiate the compact in a final effort to delay execution of further projects on the Colorado of which it does not approve. Because of the great distrust with which the other states view California, it is impossible for the western states to agree, for example, on recommendations for the all-important office of Secretary of the Interior. Any person suggested by California interest would almost automatically be opposed by the other western states and vice versa. The western states are still convinced that Ray Lyman Wilbur favored California when he was Secretary of the Interior, and in proof of this contention, point to the employment of Northcutt Ely, Assistant Secretary of the Department under Mr. Wilbur, as California's chief water lobbyist at a salary of $40,000 a year. The bitterness of feeling against California can be suggested by the leads on recent editorials in the Denver *Post:* "New Low

[4] Denver *Post,* Feb. 8, 1948.
[5] San Francisco *Chronicle,* Dec. 5, 1948.

in Trickery," [6] "California 'Harpies' and Colorado Progress"; [7] "Watch Out for California"; [8] and "Imperialism in the West: Threat to Our Water." [9] "California, and particularly Southern California," reads one of these editorials, "already looming over the West like a colossus, is determined to become the world's largest community, and is absolutely ruthless as to who else may have to pay the bill." It is this discrepancy in size, in bargaining power, that makes for troubled relations. "The lower river basin," comments Wyoming's representative, "has had all kinds of trouble *because California is so strong*—California, the Colossus of the South." [10] Precisely because of this discrepancy in power, the other western states are forming alliances against California and it is these alliances that California needs to fear more than anything else at the moment.

Southern California's policy of water imperialism in relation to the other states of the Colorado River Basin finds its almost exact counterpart in the relationship which prevails between the City of Los Angeles and other Southern California communities interested in Colorado River water. The Metropolitan Water District, originally made up of 13 communities in the Los Angeles Basin, now includes some 23 communities. In December, 1948, the cities of the Chino Basin—Ontario, Pomona, Upland, Chino, and Fontana— applied for admission to the district so as to receive Colorado River water. These communities have a combined territory of about 93,500 acres and assessed wealth of $100,500,000. Fourteen of the 23 communities making up the district voted to admit the new applicants but 4 of the 7 votes allotted to the City of Los Angeles were sufficient to defeat the application. For, in setting up the district, Los Angeles was given 50 per cent control of the board and the Los Angeles members of the board are bound by the unit rule so that a bare majority of the Los Angeles delegation can block the admission of additional communities. Thus in relation to the rest of Southern California, Los Angeles plays the

[6] Denver *Post*, Feb. 29, 1948.
[7] *Ibid.*, Dec. 31, 1948.
[8] *Ibid.*, July 6, 1947.
[9] *Ibid.*, July 7, 1947.
[10] *Ibid.*, Aug. 1, 1948.

same role that it plays, in the name of Southern California, in relation to Arizona. The tighter the water squeeze becomes, therefore, the sharper will the cleavages become between communities in Southern California. The cities of the Chino basin contend, and with good reason, that even if there were not enough water for Los Angeles, which is far from being the case, they should not be arbitrarily excluded from the charmed circle of Colorado River water users.

The matter has been well stated by Kimmis H. Hendricks.[11] Southern California, he points out, has looked far into the future in developing its water plans but it has not looked far enough to foresee the effects of the antagonism which its policy of water imperialism has engendered. In going ahead in a unilateral way with its own water plans, Southern California has never admitted "what is becoming more apparent all the time—that the Colorado belongs to a region, not a state; that its problems are regional problems, not state problems; that its beneficial use is of just as real concern to California's neighbors as to Californians themselves. Is it desirable, asks Mr. Hendricks, that Southern California should become the world's largest community? What are the risks involved in this concentration of industry and population in an area which could be paralyzed by the dropping of two bombs, one on the Owens Valley aqueduct and the other on the Colorado River aqueduct? "Hasn't the hour come," he asks, "for California, actually pre-eminent in its development of the Colorado so far, to give all possible support to regional rather than merely local planning for the ancient river's profitable use?"

The question is a good one. Actually what benefits Arizona clearly benefits Southern California. Arizona sent more than 256,-000 beef cattle to Southern California markets in 1945 (97 per cent of Arizona's shipments and one-fourth of the cattle sold on the Los Angeles markets that year). The price of Arizona alfalfa is determined by the market price at El Monte, California, with Arizona shipping 575,000 tons a year to this market. In 1945, 2,421 carloads of Arizona's choicest fruits and vegetables were shipped to Los Angeles, as well as 3,125 hogs, 61,850 sheep,

[11] *Ibid.*, March 2, 1948.

58,422 pounds of butter, and 104,511 pounds of dressed poultry. Over 200 trucks shuttle back and forth between Phoenix and Los Angeles every day. In return for these shipments, Southern California sent $25,000,000 in manufactured goods to Arizona. As Arizona points out in its water propaganda, Arizona and Southern California are one trade area; one investment market. "Like it or not," they say, "we're married." These same propagandists have coined the word "socalizona" to refer to the Southern California-Arizona unity of economic interest. Is this the penalty which one section must pay for the failure of another section to cooperate in the development of the entire region? In the long run California simply must have the cooperation of the other western states; hence its present policies are extremely short-sighted and essentially self-defeating.

On Western Unity

The problem of California, however, is not synonymous with the problem of western unity. Even if California were not a part of the West, the other western states would still face a serious problem in reconciling their claims to a limited water supply. "Water," as Richard L. Neuberger has written, "is a crucial element in the Far West" where the problem is essentially to get the water at the right place at the right time. Throughout the West, there is a remarkable discontinuity of resources both on a north-south and east-west basis. In almost every western state there is a serious cleavage between watershed and non-watershed users. Divided by the Rocky Mountains, Colorado has found it necessary to divert 24,294 acre-feet of water annually from the Fraser River, on the western slope, through the Moffatt Tunnel to Denver and other communities on the eastern slope. The Colorado-Big Thompson project involves a similar diversion of 310,000 acre-feet from one watershed to another, not to mention the Arkansas project by which 800,000 acre-feet of water from the Gunnison River is transferred from the western to the eastern slope. It is not surprising, therefore, to find that 18 western slope counties have banded to-

[12] *Ibid.*, May 14, 1949.

gether to prevent further water diversion from the western to the eastern slope. Even in the Pacific Northwest, there is sharp controversy within the region over the control of water.[13] In state after state the same disunity appears, based on conflicting uses or geographical divisions.

The fact is, of course, that the establishment of state boundaries in the West was determined, not by local needs, but by the necessities of national partisan politics. Idaho, for example, is a geographical monstrosity, being made up of three regions each of which is cut off from the other and dependent upon different economic interests. Some of these states were admitted before they had an adequate population base; others long after their resources had been monopolized. Nevada was admitted *too* soon (1864); Arizona and New Mexco *too* late (1912). The issue of polygamy long delayed the admission of Utah, although Utah had a larger population and was better able to administer its own affairs than states which had been admitted at a much earlier date. The constitution of Nevada was telegraphed to Washington so that the state might be hastily admitted. In this case, the Republican Party needed additional votes to adopt the 13th Amendment. Montana, Idaho, and Washington were also hastily and improvidently admitted to insure the ascendancy of the Republican Party. Artificial boundaries, often the result of improvised party politics, have imposed a heavy burden on the western states in the way of highway construction, the maintenance of social services, the laying out of communications, and, above all, in the development and utilization of water resources.

In the process of being admitted to the Union, particular handicaps were imposed on certain western states. For example, when Arizona was finally admitted to the Union, Congress insisted that the constitution of the new state should contain a provision by which Arizona agreed that the United States might withdraw from entry and reserve all of the power dam sites on the Colorado River across the state with the right to withdraw and reserve the lands bordering that stream in Arizona. Thus Arizona never had

[13] "To the West, Water Is Life and Death," by Richard L. Neuberger, *New York Times* Magazine October 24, 1948.

the ordinary rights enjoyed by the other basin states to control or
to build or operate dams and diversion works from the Colorado;
in effect, the state was handcuffed. Prior to statehood, moreover,
Arizona, as a territory, had to remain idle while Imperial Valley
staked out claims on Colorado River water. In fact, the singular
provision in Arizona's constitution to which I have referred, so it is
said, was drawn up by Imperial Valley interests for the purpose
of preventing Arizona from developing uses in the Colorado River.
At the time the Arizona constitution was adopted, it should be recalled, Mr. Harry Chandler, publisher of the Los Angeles *Times*,
owned 833,000 acres of land in Mexico immediately below the
border of which some 600,000 acres were irrigable from the water
of the Colorado River.

The discontinuity of resources in the West is matched by the
social discontinuity and historical unrelatedness of the region.
Communications are of vital importance in a region of such vast
distances. Yet the rail and highway traffic arteries run east and
west, not north and south, despite the fact that the mountain
ranges happen to run on a north-south line. The fact is, of course,
that the West's communications were designed to serve national
rather than regional interests. This consideration alone has made it
extremely difficult for these states to develop a sense of internal
unity or a sense of unity within the region as a whole. The western
wire-services and news-gathering agencies are also organized on an
east-west basis and in terms of national interest rather than regional convenience. Within the West, as a region, there is an
amazing lack of intra-regional exchange and communication. From
the standpoint of news, Boise is as remote from Los Angeles as
Raleigh, North Carolina. On the map the western states are
neighbors; but in point of social fact they are discontinuous entities, separated by mountain ranges, canyons, miles of desolate terrain, and, above all, by impossible communications. Many of the
western communities, as Dr. Carle C. Zimmerman has pointed out,
are "oasis-like" settlements which tend to look inward rather than
outward since the oases are "intense regionalizations of diverse
cultures." As Dr. Zimmerman states, "It takes a bishop to get anywhere in Utah and a padrone in New Mexico." Within these

states, also, there is a remarkable historical discontinuity. California's historical development, for example, does not fit the pattern of the other western states, each of which has its own particular tradition or myth. Today there is not a single publication which attempts to relate or interpret the interests of the West as a whole. All of these factors must be considered in order to understand the problem which now faces these states, that is, the problem not of agreeing upon a resource development plan but, first, of agreeing-to-agree.

Consider, for example, some of the political problems which the matter of water conservation and development presents. Where water is involved, western politicians cannot compromise; they dare not yield an acre-foot of water. In the Tri-State Commission meeting in 1940, California, Nevada, and Arizona had in effect reached an agreement on the division of Colorado River water; but, as one negotiator said, "the rock we broke on was the question of responsibility." In other words, who was to announce the agreement? who was willing to accept responsibility for its negotiation? None of the negotiators were willing to admit that he had made any concessions. The businessmen in these states, who understand regional markets, do not understand water technology. And, to further complicate matters, each state has a collection of highly paid "water experts" who have a vested interest in their jobs. The "professional water men," in fact, are a major problem in themselves. To create an atmosphere in which discussion is possible, most interstate water conferences meet in secret session with the press excluded, for none of the negotiators wants to be quoted. The secrecy of the meetings in turn prevents the people of the various states from understanding the facts and makes them peculiarly vulnerable to the "power politics" cooked up by the experts and peculiarly dependent upon these experts for factual information. Every state has its particular "water secrets" which it is unwilling to disclose, so that the full facts about any particular controversy are seldom revealed.

The current fashion, of course, is to blame California for the failure to achieve a real western unity. Among the western states, writes Richard Steinbruner, "California is the biggest boy in the crowd, and has possession of more of the marbles. The other boys, shouting that California is a bully, are willing to gang up to see that California doesn't get hold of any more water." [14] In the editorial comment of the other states, California is invariably described as "big, rich, and clever." Although there is a great deal to this charge, as I have tried to explain, the other western states fail to recognize that it is inherently difficult for California *to relate* its interests and problems to those of the West as a whole. Many of California's problems have utterly no relevance to the other western states. Aside from California, what other western state is interested in the problem of marketing oranges, tangerines, avocados, raisins, grapes, figs, pomegranates, and lemons? California has long been forced to concentrate on its own problems for the simple reason that these problems are of no concern to *any* state other than California. California has been forced, therefore, to develop the techniques for coping with these problems and to push through whatever federal legislation was needed. Being of little concern to other states, the measures which California has requested have, in the past, encountered little opposition and the federal government has permitted California, so to speak, to write its own ticket. Conversely, many western problems are of little concern to California as, for example, Indian lands, grazing rights in the forests, public lands generally; even California's sheep and cattle industries are "exceptional." The inter-mountain states are not interested in fisheries, foreign trade, migration; they are concerned with the livestock industry, domestic trade, and the tourist business. And so it goes. Not only has there been remarkably little cultural exchange between these states but they have had virtually no experience in working together politically on issues. Political movements which have swept California have, for the most part, originated within the state and have had little repercussion outside the state.

It isn't simply California's superior power which complicates western unity, but the fact that there is little unity within the West on any issue. Assuming that California wanted to cooperate with the other western states and that these states wanted to cooperate

[14] Denver *Post*, March 25, 1948.

with California, the question would still remain: what do these other states cooperate with? with what particular California should they seek to establish some common bond of interest? For the fact is that California has not yet achieved the internal, the organic unity which would make cooperation and better relations with the other western states possible. The California Congressional delegation is more likely than not to be broken up into rival factions on most issues. With which faction should the other western states cooperate? California is a political cauldron; seldom, if ever, is a governor re-elected. California is Northern California vs. Southern California; newcomers vs. carpetbaggers; hinterland vs. metropolitan areas. It is, as the Los Angeles *Daily News* pointed out in an editorial, "an unarticulated collection of factors looking for a common denominator." [15] And the point made by this editorial, namely, that "California hasn't yet the kind of unity it must have if it is to win the assistance and encouragement of other sections and states and of the nation as a whole," comes close to being the essence of the California problem. When the other western states complain, therefore, that "they are sticking together—all except California," they forget that California, as the largest and most mature western state, is suffering acutely from a disease which is common throughout the region, namely: a lack of social and cultural integration.

THE FUTURE IS WEST

Reduced to its essentials, *the California Problem* might be described as follows: a large province on the west coast, occupying a marginal geographical position, possessed of a most exceptional environment with its peculiar advantages and no less peculiar problems, with a most unique historical background, gets a head-start of two decades over the other western states in its development and, because of a set of exceptional internal dynamics, develops at an entirely different tempo from its sister states of the West. The uniqueness and novelty of the environment, coupled with its amazing versatility, operates as a constant challenge to

[15] Los Angeles *Daily News*, May 7, 1948.

social and technological inventiveness. Selective forces at work in the process of migration bring to the state a population which is not so much a cross-section as a highly selected sample of the population of the world. The diversity of the state's resources is matched by the constant diversity of its population. Preoccupied with its peculiar problems, isolated from the rest of the nation during two crucial decades in its early history, California develops a remarkable energy and resourcefulness in the solution of its problems without consultation or assistance from the other western states or the federal government.

Over a period of years, therefore, a spirit of great independence and a self-reliance bordering on truculence develops among its people. But the circumstances which have shaped and moulded this far west province have been such that its people have tended to ignore the fact that, despite its marginal position, or precisely because of this position, the state is closely related to and dependent upon a larger region. The discrepancy in power, as between California and the other western states, has strikingly augmented this tendency on the part of the people to think of California as a province apart, sovereign in its own right, a self-contained empire. The very scale by which happenings, events, and developments are measured in the freakish environment of California makes it extremely difficult for Californians *to relate* their problems to those of the other western states. The environment, in other words, tends to distort the perspective of those who dwell within the state. The scale is so much larger; the tempo of events so much faster; and, in California, everything seems to be reversed, to occur out of the natural order of events, to be upside down or lopsided. Even to describe the state accurately is to run the risk of being branded a liar or a lunatic. Add to these considerations the fact that California is still *new*, almost as new as it was in 1848, and one has a pretty good idea of what makes up the California problem.

How is this problem to be resolved? In the first place, events are rapidly resolving it. This west coast "panther with the splendid hide" is being tamed and made to know its master's will by the inescapable logic of events. For great and diverse as its resources are, they are still inadequate to meet the needs of the population of 20,000,000 people that will one day be residing within its border. Forced to seek federal assistance on a large scale, California will be brought face to face with the fact that the federal government must intervene not at one but at many points in its internal affairs. Coming to a new life, the other western states are now arrayed against California on many issues and California will be driven to seek the cooperation of these states whether it likes the idea or not.

But, more potent than events of this character, is the fact that California's destiny, which can be perceived but dimly today, will correct the balance by investing California, willy-nilly, with the role of western leadership. Events of a magnitude too vast even for conjecture are taking place today around the rim of the Pacific: in China, India, the Philippines, Java, Sumatra, French Indo-China, the Soviet Far East. Regardless of how these events work out, one thing is certain: California is destined to occupy in the future, not a marginal, but a central position in world affairs. The ports of the west coast will be the ports through which the expanding trade and commerce of the West will flow to ports throughout the entire vast area of the Pacific. Once the impact of this development really begins to make itself felt, California will come to occupy a new position in the western scheme of things; not that of the Colossus of the West, the Big Bully, the Untamed Panther, but the state which will link western America with the Orient.

Possession of Alaska and the Aleutian Islands gives the United States a Pacific base which extends for 4,000 miles from San Diego to Attu and is only 600 miles, at the nearest point, from the islands of Japan. Historic areas of civilization, Ellen Churchill Semple pointed out, are most naturally indicated by the seas which they encompass. When the shift came from the Mediterranean to the Atlantic, the nations which had a front seat on the Atlantic naturally came to dominate the world. In fact, Miss Semple thought that the pre-eminence which these states have long enjoyed was so great that the initiative would never pass completely to the larger basin of the Pacific. But, despite this reservation, Miss Semple was convinced that the Pacific was "the ocean of the future" since it represents "the final expansion of the maritime field" and the "sea is always one." In the past the demands of the Orient have been those of old, crowded countries which had not advanced into the modern industrial stage of development; but, as these demands multiply, as they will, the west coast is certain to be galvanized into a new type of industrial activity. Developments which now appear to be of doubtful long-range value may come to assume an enormous importance in the future. Today there are those who believe that Southern California is already greatly over-industrialized in terms of its resource-base; but this concentration of industry may be regarded in an entirely different light two decades hence. The Pacific as "the ocean of the future" is still merely an oratorical phrase, a rhetorical flourish, a theme for chamber of commerce bombast; but west coast industrialists who are already beginning to fill orders from India are convinced that the oratorical phrase of yesterday may become the economic reality of tomorrow.

The problem of the West, therefore, is to build toward the future; toward the Pacific. There are all too many indications, however, that the development of the West is being predicated on fragmentary information, improvised planning, and the opportunistic promotion of "projects." Confusion and uncertainty prevail at the moment in almost every aspect and phase of this development. Do we need a Columbia River Authority? a Central Valley Authority? a Colorado River Authority? Should we have an overall board of review to analyze all river basin developments and to coordinate these developments? Should the federal government spend billions to provide water to states that refuse to adopt a code regulating the right to pump underground waters? Should the development of the West be determined by the relative strength of various pressure groups? Should this development be planned by agencies which have a "vested interest" in constructing projects as projects? What sort of socio-economic policies should govern these gigantic development schemes? It is doubtful if the answers to these and many similar questions will come from the wrangling of congressional debates or the feuding between rival federal agencies or the power politics of states. The first thing that is needed is regional fact-finding on a sufficiently comprehensive basis to provide the foundation for a long-range development program. Once gathered and correlated, these facts should be presented directly to the people, as through a series of western states conferences, so as to give the people some direct knowledge of the interrelation of western resources and the mutual dependencies which prevail between these states.

On the state capitol at Sacramento one can read the scroll: "Bring Me Men to Match My Mountains!" This is California's need today: for men and women who can match, in the scale of their imagination and the depth of their insight, the extraordinary diversity, power, and challenge which is implicit in this immense and fabulous province which sprawls along the Pacific like a tawny tiger. California needs men who can see beyond its mountains; men who can see the entire West and who realize that, as with all good things, there comes a time when the gold runs out, when the exception disappears in the rule, and when California "being so caught up, so mastered by the brute blood of the air" must, indeed, put on knowledge with its power and adopt, as an official policy, the same generous open-handedness with which its magic mountains have showered benefits on those lucky people, the Californians.

ACKNOWLEDGMENTS

◇◇◇

I WISH to acknowledge my indebtedness to the following individuals for various courtesies and helpful suggestions: Oscar Fuss, A. C. Prendergast, J. Rupert Mason, Dr. James J. Parsons, Dr. Eshref Shevky, Dr. Eric Temple Bell, Max Stern, Richard Boke, Dr. Paul S. Taylor, Robert W. Kenny, Margaret O'Connor, John Anson Ford, Charles A. Carson, Mildred Edie Brady, and Dr. Samuel B. Morris.

I would also like to pay tribute to the following sources which I found to be particularly helpful: *The Population of California,* a report prepared in 1946 for the Commonwealth Club of California by Dr. Davis McEntire; *San Francisco: Port of Gold,* by William Martin Camp (Doubleday & Co. 1947); *The Gold Rushes* by W. P. Murrell (MacMillan Co. 1941); *The Colorado Conquest* and *The Glass Giant of Palomar,* both by David O. Woodbury and published in 1941 and 1946, respectively, by Dodd, Mead & Company; *Men, Mirrors and Stars* by G. Edward Pendray, Funk and Wagnalls, 1935; *Gold is The Cornerstone* by Dr. John Walton Caughey, University of California Press; *California Gold* by Dr. Rodman W. Paul, Harvard University Press, 1947; *San Francisco's Literary Frontier* by Franklin Walker, Alfred A. Knopf, 1939; *California Agriculture,* edited by Claude B. Hutchison, University of California Press, 1946.

Portions of this manuscript have appeared in *Harper's* and *The Nation* and are reprinted here with the permission of the editors.

INDEX

Index

[*371*]

Index